Summer Counts!

2nd Edition

Grade 5→6

Thematic Reading, Language Arts, and Math Skills

Options™

ISBN 978-1-60161-930-3

OT105

Cover Image: Enrique Corts/Shannon Associates, LLC

Triumph Learning® 136 Madison Avenue, 7th Floor, New York, NY 10016

© 2010 Triumph Learning, LLC
Options is an imprint of Triumph Learning®

All rights reserved. No part of this publication may be reproduced in whole or in part, stored in a retrieval system, or transmitted in any form or by any means, electronic, mechanical, photocopying, recording or otherwise, without written permission from the publisher.

Printed in the United States of America.

10 9 8 7 6 5 4 3 2 1

Dear Parent,

Summer is a time for relaxing and having fun. It can also be a time for learning. *Summer Counts!* can help improve your child's understanding of important skills learned in the past school year while preparing him or her for the year ahead.

Summer Counts! provides grade-appropriate practice in subjects such as reading, language arts, vocabulary, and math. The ten theme-related chapters include activities and puzzles to motivate your child throughout the summer.

When working through the book, encourage your child to share his or her learning with you. You may want to tear out the answer key at the back of the book and use it to check your child's progress. With *Summer Counts!* your child will discover that learning happens anytime—even in the summer!

Apreciados padre,

El verano es una época para descansar y divertirse. También puede ser una época para aprender. *Summer Counts!* puede ayudar a que su hijo(a) mejore las destrezas importantes que aprendió el pasado año escolar al mismo tiempo que lo(a) prepara para el año que se aproxima.

Summer Counts! provee la práctica apropiada para cada grado en las asignaturas como la lectura, las artes del lenguaje y las matemáticas. Los diez capítulos temáticos incluyen actividades y rompecabezas que motivarán a su hijo(a) durante el verano.

Cuando trabaje con el libro, anime a su hijo(a) a que comparta lo que ha aprendido con Ud. Si Ud. desea puede desprender la página de las respuestas que aparece en la parte trasera del libro. Puede usar la misma para revisar el progreso de su hijo(a). ¡Con *Summer Counts!* su hijo(a) descubrirá que el aprendizaje puede ocurrir en cualquier momento—inclusive en el verano!

Table of Contents

Letter 2

Chapter 1: Family
An African Folktale: The Princess Twins 4
Compound Nouns 6
Family Scramble 7
Like and Unlike Fractions 8
Money Fractions 9
Relative Riddles 10
Challenge Yourself: Reading 11

Chapter 2: Occupations
Faces in Stone 12
Predicate Nouns 14
Job Analogies 15
Mixed Numbers 16
Painting Problems 17
People Puzzle 18
Challenge Yourself: Math 19

Chapter 3: Places
Reach for the Sky 20
Pronouns 22
Vacation Variation 23
Estimating Mixed Numbers 24
Pool Proportions 25
Around a Building 26
Challenge Yourself: Math 27

Chapter 4: Sports
Extraordinary Athletes 28
Verb Phrases 30
Sports Shorts 31
Adding Mixed Numbers 32
Hiking for Miles 33
Secret Sports 34
Challenge Yourself: Reading 35

Chapter 5: Music
The Spirit of Jazz 36
Verb Tenses: Past, Present, and Future 38
Same Songs 39
Subtracting Mixed Numbers 40
Penny Probability 41
Song Search 42
Challenge Yourself: Math 43

Chapter 6: Animals
Why Animals Can't Talk 44
Predicate Adjectives 46
Wildlife Comparisons 47
Working with Decimals 48
Measure the Distance 49
Animal Imposters 50
Challenge Yourself: Reading 51

Chapter 7: Ocean
Scotland's Sea Monster 52
Conjunctions 54
Landscape Likeness 55
In the Hundredths Place 56
Fish Finds 57
Sea Scramble 58
Challenge Yourself: Math 59

Chapter 8: Holidays
Bai-Nien 60
Subjects: Simple and Complete 62
Holiday Picnic 63
Comparing Decimals 64
Track Times 65
Holiday Puzzle 66
Challenge Yourself: Reading 67

Chapter 9: Students
A New Student 68
Sentence Predicates 70
Time Scramble 71
Adding Numbers with Decimals 72
Number Sense 73
Student Search 74
Challenge Yourself: Math 75

Chapter 10: Plants
The Plant Doctor 76
Sentences 78
Pick Your Plants 79
Subtracting Decimals 80
Cube Creations 81
Plant Codes 82
Challenge Yourself: Reading 83

Reading Check-Up 84
Math Check-Up 88
Certificate of Completion 91
Answer Key 92

CHAPTER 1
An African Folktale: The Princess Twins

Long ago in Africa, two beautiful twin sisters ran and played together. Their names were Zinsa and Zinhoue. They looked exactly alike except that Zinsa was born with a shiny silver bracelet on her wrist, while Zinhoue had shimmering feathers in her hair.

The twins were so lively and beautiful. Everyone loved the sisters, except for the two old women of the forest. They each wanted the bracelet and feathers for themselves. The old women thought the bracelet and feathers would make them look young again.

One day, Zinhoue became very sick. She heard the old women calling her down to the Forest of Bliss. "Zinhoue," whispered her sister, "don't leave me alone. What do the old women want?" They wanted Zinsa's shiny silver bracelet.

Zinsa thought giving them the bracelet might make Zinhoue better, so she cut the bracelet off her wrist. She gave it to the old women. Immediately, Zinhoue woke up, but Zinsa had no adornments and was seen as plain by many people.

When the girls were older, two princes entered their kingdom. The princes had never seen twins before and wanted to marry Zinsa and Zinhoue. However, the law of the land said that princes had to marry beautiful girls with many riches and adornments. Zinsa was now seen as dull without her bracelet. The princes told the girls not to worry and dressed them in beautiful, fancy robes to hide Zinsa's bare wrist.

Everyone in the new kingdom loved the girls, except for the princes' sister, Althea. She became very jealous of them. One day, Althea saw that Zinsa had no adornments or decorations. She was just as plain as the other townspeople!

Althea wanted everyone to know this. She thought of a way for everyone to find out. She planned a contest in which all the girls in the town would compete to see who could grind the most millet. Zinsa was afraid. She knew she could not grind millet while wearing her fancy robe. Everyone would see how plain she was!

Zinhoue had a plan. The girls dressed in their hooded robes. Zinhoue and Zinsa went first. Everyone watched as the twins tried to grind the millet in their robes. Suddenly, Zinsa took off her robe to reveal her bare, plain wrists. The crowd gasped! When Zinhoue stood up to protect her sister, her hood fell back to reveal plain, short hair. All her shimmering feathers were gone! She, too, was as plain as a commoner.

Just then, birds flew from the sky and cracked the millet with their beaks. They ground more millet than anyone else did. The girls won!

"Zinsa once gave her beauty and riches to save my life. Now I have given up mine to save hers," said Zinhoue. Everyone cheered. After that, the townspeople changed the law to let princes marry anyone they wanted, adorned or plain.

An African Folktale: The Princess Twins

Directions Using what you have just read, answer the questions.

1. How are Zinsa and Zinhoue the same? How are they different?
   _____
   _____

2. Why did Zinsa cut off her bracelet?
   _____
   _____

3. The princes marry the sisters even though there is a law against it. How do you think the princes feel about the law?
   _____
   _____

4. Why was Althea jealous of the twins?
   _____

5. Zinsa and Zinhoue save each other. What does this say about them?
   _____
   _____

6. Why do you think the townspeople changed the law to allow princes to marry anyone they wanted?
   _____
   _____

Compound Nouns

REMEMBER
A **noun** is a word that names a person, place, or thing.
A **compound noun** is a noun that is formed from two or more words.

EXAMPLES

Noun	Compound noun
brother	brother-in-law
kitchen	living room
apple	grapefruit

Compound nouns can be written in three ways:

one word	two words	words with a hyphen
grapefruit	living room	great-aunt

Directions Draw a line under the compound noun or nouns in each sentence. Then write the words that make up the compound nouns on the lines.

1. Tim's grandparents live in Missouri. _____ _____

2. Their house is near a golf course. _____ _____

3. Tim's grandfather, Elias Beech, owned a newspaper.

   _____ _____ _____ _____

4. Mr. Beech worked for the paper as a teenager. _____ _____

5. When Mr. Beech retired, he bought a bookstore. _____ _____

6. The store is next to the post office. _____ _____

7. Mr. Beech's sister-in-law works at the store.

   _____ _____ _____

8. She taught English at the high school. _____ _____

9. Tim likes exploring the bookshelves at the store. _____ _____

10. He reads adventure stories and science fiction. _____ _____

Language Grade 5

Family Scramble

Directions Unscramble the letters to make words from the box.

> nephew   brother   cousin
> relative   parents   sister

1. ivealert _____
2. siucon _____
3. btorhre _____
4. whepen _____
5. ssreit _____
6. snarpte _____

The Family Tree

Directions Read the conversation. Use words from the box below to fill in the blanks.

> brother   grandmother
> marriage   grandchildren

Vince: "Grandma, could you please tell me about our family?"

Grandmother: "Sure. You know, my mother, your **(7)** _____, came over to America from Italy when she was a little girl. She came with her parents—my grandparents—and her **(8)** _____, who is my uncle, Paul."

Vince: "He's the one who always calls you granny, right?"

Grandmother (laughing): "Yes, that's right. He had four children, and I had three, including your mother. His third son, Sam Jr., married Kathy, who had two boys from a previous **(9)** _____. We welcomed them into our family right away."

Vince: "Wow, I didn't realize we had such a big family!"

Grandmother: "Yes, and with you and the other twelve **(10)** _____ it just keeps getting bigger!"

Like and Unlike Fractions

REMEMBER
To add or subtract like fractions, use the same denominator.
To add or subtract unlike fractions, find a common denominator.

EXAMPLES

Add like fractions

$$\begin{array}{r} \frac{2}{4} \\ +\frac{1}{4} \\ \hline \frac{3}{4} \end{array}$$

Add unlike fractions

$$\begin{array}{r} \frac{3}{8} = \frac{3}{8} \\ +\frac{1}{4} = \frac{2}{8} \\ \hline \frac{5}{8} \end{array}$$

Subtract unlike fractions

$$\begin{array}{r} \frac{3}{4} = \frac{9}{12} \\ -\frac{2}{3} = \frac{8}{12} \\ \hline \frac{1}{12} \end{array}$$

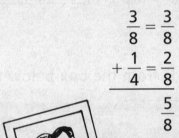
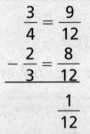

Directions Add or subtract the like and unlike fractions. Reduce to lowest terms.

1. $\frac{2}{4} + \frac{1}{4}$

2. $\frac{5}{9} + \frac{1}{9}$

3. $\frac{5}{6} - \frac{3}{6}$

4. $1 - \frac{4}{10}$

5. $\frac{2}{3} = \frac{\square}{6}$
   $+\frac{1}{6} = \frac{1}{6}$

6. $\frac{5}{8} = \frac{5}{8}$
   $+\frac{1}{4} = \frac{\square}{8}$

7. $\frac{2}{3} = \frac{\square}{6}$
   $-\frac{3}{6} = \frac{3}{6}$

8. $\frac{7}{10} = \frac{7}{10}$
   $-\frac{3}{5} = \frac{\square}{10}$

8 Math *Grade 5*

Money Fractions

Directions Complete the page.

1. A coin can be thought of as a fraction of a dollar. For example, 1 penny is $\frac{1}{100}$ of a dollar because there are 100 pennies in one dollar.

   What fraction of a dollar is each coin below?

   A.  quarter _____

   B. dime _____

   C. nickel _____

2. A. Write the following equation in numbers. Then solve to find the total value.

   $\frac{1}{4}$ of a dollar + $\frac{1}{10}$ of a dollar + $\frac{1}{20}$ of a dollar =

   _____ + _____ + _____ = _____

   B. Now write the answer as a fraction of a dollar. Reduce the fraction to lowest terms. _____

3. Find the sum. Reduce the answer to lowest terms. Show your work.

   $\frac{1}{4} + \frac{3}{10} + \frac{7}{20} =$ _____

Math Grade 5

Relative Riddles

Directions Read each clue. Name the family member described to solve the riddle.

1. I am your mother's sister. Who am I?
   _____

2. I am your father's mother. Who am I?
   _____

3. I am your mother's son. Who am I?
   _____

4. I am your mother's father. Who am I?
   _____

5. I am your aunt's son. Who am I?
   _____

6. I am your father's brother. Who am I?
   _____

7. I am your mother's daughter. Who am I?
   _____

8. I am your mother and father. Who am I?
   _____

Challenge Yourself: Reading

Oh Brother!

Jan had been trying to finish her painting all week, but her little brother, Carlos, kept interrupting her. He wanted her to come out and play with him. So when Jan heard her brother knocking on her bedroom door, she shouted, "WHAT IS IT?"

Although she was irritated, Jan wiped her hands on a paint-spotted towel and opened the door. She looked out and saw her brother sneaking away down the hallway. Then she noticed something on the floor. It was a bowl of her favorite cereal and a glass of orange juice, arranged nicely on a tray.

Jan smiled as she picked up the breakfast tray. As she ate, she looked at her painting. It was a picture of Carlos playing soccer. It was almost finished, and she couldn't wait to show it to him.

Directions Circle the letter of the **best** answer.

1. Where does this story take place?

    A. a soccer field   **B.** a kitchen

    C. Jan's bedroom   **D.** an art gallery

2. Why do you think Jan wiped her hands on a towel before she opened the door?

    A. She had just washed her hands.

    B. She didn't want to spread germs.

    C. She had paint on her hands.

    D. She had spilled orange juice on her hands.

3. Why does Carlos keep interrupting Jan?

    A. He misses playing with Jan.

    B. He thinks painting is a waste of time.

    C. He wants to annoy her.

    D. He is excited about Jan's painting.

4. Which word from the story is a compound noun?

    A. hungry   **B.** bedroom

    C. picture   **D.** playing

Challenge Yourself: Reading Grade 5

CHAPTER 2
FACES IN STONE

Gutzon Borglum (1867–1941) was a sculptor. He had big ideas about art. Borglum didn't paint, though. He carved in stone. His love of art was matched only by his love of America and its history. He believed that America's story should be told through great art. It was his love of art and his country that led his eyes upward—to the mountains. When Borglum looked up at a mountain, he didn't see just a rocky peak, but instead a piece of stone waiting to be carved into a work of art.

In 1923, a man named Doane Robinson proposed that a great piece of art be carved into the Black Hills of South Dakota. Robinson asked Borglum to think about the idea. The carving would be a memorial of people who were important to South Dakota's history, such as Lewis and Clark, and Kit Carson.

Borglum wanted the job, but he did not like the subjects chosen. For such a large piece of art, he thought four great presidents of the United States would be a better subject. With this carving, he wanted to honor George Washington, Thomas Jefferson, Abraham Lincoln, and Theodore Roosevelt. He also proposed carving into Mount Rushmore because its 5,725-foot-high surface was smooth and hard. People liked Borglum's idea, and in 1925, Mount Rushmore was set aside as a National Memorial.

Borglum spent a lot of time designing the model of the four presidents. On August 10, 1927, a team of over 100 men began the work of turning Borglum's model into a huge stone carving. The men spent their days hanging over the side of the mountain, sculpting faces that were about 60 feet tall. Until 1941, Americans watched as, one by one, the presidents of Mount Rushmore appeared in stone. When it was finally done, Borglum had sculpted a masterpiece of American art.

Faces in Stone

Directions Using what you have just read, answer the questions.

1. Who was Gutzon Borglum?

   _____
   _____

2. What two things did Borglum love?

   _____
   _____

3. Why do you think Borglum did not like the subject Doane Robinson suggested for the carving?

   _____
   _____

4. What happened after Borglum made the model of the four presidents?

   _____
   _____

5. The faces of which four presidents were carved into Mount Rushmore?

   _____
   _____

6. How long did it take to carve the faces into Mount Rushmore?

   _____

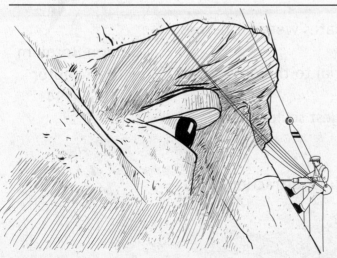

Reading Comprehension *Grade 5*

Predicate Nouns

REMEMBER
A **predicate noun** tells what a subject is or is like. It follows linking verbs, such as **am, is, are, was, were, being,** and **been**.

EXAMPLES
Mount Rushmore is a famous **mountain** in South Dakota.
Gutzon Borglum was an **artist**.

Directions Read each sentence. Draw a line under the predicate noun.

1. Mount Rushmore is also a national monument.

2. The monument is a huge carving.

3. The carving is a portrait of four presidents' faces.

4. Gutzon Borglum is the artist who created the sculpture.

5. Borglum was the son of Danish parents.

6. He was a sculptor in Idaho.

7. The presidents of the United States were his heroes.

8. Borglum's sculpture is a memorial to the presidents.

9. The memorial is one of the largest sculptures in the world.

10. The monument is a popular tourist attraction.

Gutzon Borglum,
the Sculptor
1867–1941

Job Analogies

Directions Analogies are made from pairs of words that have the same relationship. Complete the analogies using words from the box below.

singer	captain	director	doctor	sculptor
governor	police	act	artist	students

1. CARPENTER is to SAW as _____ is to STETHOSCOPE.

2. PUBLISHER is to BOOK as _____ is to MOVIE.

3. BAND is to _____ as MOVIE is to ACTOR.

4. OFFICE is to EMPLOYEES as CLASSROOM is to _____.

5. PRESIDENT is to COUNTRY as _____ is to STATE.

6. _____ is to PAINT as _____ is to STONE.

Career Goals

Directions Read the selection. Use words from the box above to fill in the blanks.

Many students have ideas about what they would like to do when they grow up. They think about activities they enjoy and special talents they have. For example, Jan is the (7) _____ of her softball team. She enjoys helping her teammates and making sure they follow the rules. She is a good leader. Someday Jan might want to be a (8) _____ officer and help people in the community.

Louis likes to act in plays. He takes acting lessons and dance lessons. He goes to the movies often. He especially likes movies made by the (9) _____ Steven Spielberg. Someday, Louis hopes to (10) _____ in a movie. It would be great to see him on the big screen!

Mixed Numbers

REMEMBER
A **mixed number** is a whole number plus a fraction.

EXAMPLES

Mixed number	Meaning	Word name	Picture
$2\frac{1}{2}$	$2 + \frac{1}{2}$	two and one-half	
$4\frac{2}{3}$	$4 + \frac{2}{3}$	four and two-thirds	

Mixing It Up

Directions Write the mixed number and word name for each shaded group.

1.

   A. mixed number _____

   B. word name _____

   _____

2.

   A. mixed number _____

   B. word name _____

   _____

3.

   A. mixed number _____

   B. word name _____

   _____

4.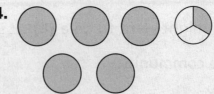

   A. mixed number _____

   B. word name _____

   _____

16 Math Grade 5

Painting Problems

Directions Read the paragraph below. Use the information to answer the questions.

Janessa is painting her room her own special color. To make the color, she mixes $\frac{3}{4}$ cup of yellow paint and $\frac{3}{8}$ cup of red paint into 1 quart of white paint. After painting $\frac{1}{4}$ of her room, she is out of paint. She wants to paint the rest of the room the same color.

1. How much more white paint does Janessa need?

   _____

2. How much more of each paint does Janessa need? Show your work.

   A. yellow paint: _____ cups

   B. red paint: _____ cups

3. What is the total amount of paint that Janessa will use for the entire room? Show your work.

   A. white paint: _____ quarts

   B. yellow paint: _____ cups

   C. red paint: _____ cups

4. Janessa decides to add a second coat of the same paint to make sure the walls are covered. How much of each paint color will she have used in all?

   A. white paint: _____ quarts

   B. yellow paint: _____ cups

   C. red paint: _____ cups

People Puzzle

Directions Read each clue. Use words from this chapter to solve the puzzle.

actor   doctor   teacher   dentist   sculptor
police   painter   artist   director

ACROSS
2. This person helps students learn in school.
5. A _____ officer helps people stay safe.
6. Someone who makes works of art
7. Gutzon Borglum was a _____.
9. A star in a movie

DOWN
1. This person looks at your teeth.
3. A movie-maker
4. Someone who paints
8. This person helps you when you are sick.

Challenge Yourself: Math

Chocolate Chip Cookie Recipe

Directions Henry is baking cookies for a school bake sale. He is using the following recipe. The recipe makes 25 cookies.

$\frac{3}{4}$ cup sugar

1 cup brown sugar

1 cup butter

2 eggs

1 teaspoon vanilla extract

$2\frac{1}{4}$ cups flour

1 teaspoon baking soda

1 teaspoon salt

$2\frac{1}{2}$ cups chocolate chips

1. Henry has $1\frac{1}{2}$ cups of flour. How many more cups of flour does he need to make one batch of cookies?

    _____

2. If Henry wants to make 100 cookies, how many cups of brown sugar will he need?

    _____

Directions Henry bought 15 cups of chocolate chips. He plans to make as many cookies as he can with these chocolate chips. Use this information to answer questions 3 and 4.

3. How many cookies can he make?

    _____

4. List the quantities of the other ingredients he will need to make these cookies.

    sugar _____   vanilla extract _____

    brown sugar _____   flour _____

    butter _____   baking soda _____

    eggs _____   salt _____

Challenge Yourself: Math Grade 5

CHAPTER 3
Reach for the Sky

In the early 1900s, many people were moving to New York City to find jobs. Everyone wanted offices and apartments in the city, but there just wasn't enough space. Workers had to move quickly to come up with a solution. New steel building methods brought that solution—build taller and taller buildings.

In 1903 the first New York skyscraper, the Flatiron Building, went up. By 1929 there were almost 200 skyscrapers filling the city skyline. Despite having little money during the Great Depression, a group of New York businessmen combined their funds to build the tallest building in the world—the Empire State Building.

At the time, the Chrysler Building was the tallest building. The builders of the Empire State Building also had another challenge. They would have to obey New York City building codes, or laws. The laws said that as a building got taller, it had to get narrower. While these laws made sure that all skyscrapers would have enough light and fresh air, they also caused some problems. As the top floors got narrower, there was less space for an elevator shaft or offices.

A team of owners, architects, engineers, and builders worked on the Empire State Building. Construction problems were solved in record time, and soon the building was well under way. The building took only one year and forty-five days to build.

The outside of the Empire State Building was made of limestone and granite with stainless-steel trim. Inside, it had 73 elevators. Best of all, it was taller than the Chrysler Building—a full 1,250 feet tall, or 102 stories.

For more than 40 years, the Empire State Building was the tallest building in New York City. It was the tallest building in the world until 1974 when the Sears Tower in Chicago was built, standing 204 feet taller.

The Empire State Building held the title of tallest building for more than 40 years. Many tourists and New Yorkers visit it each year. The Empire State Building may no longer be the tallest in the world, but it's tall enough!

Reach for the Sky

Directions Using what you have just read, answer the questions.

1. What is this article mostly about?
   _____
   _____

2. What challenges did the builders of the Empire State Building face?
   _____
   _____

3. What are building codes and why are they important?
   _____
   _____

4. How long did it take to build the Empire State Building?
   _____

5. What skyscraper was built after the Empire State Building?
   _____

6. How do skyscrapers built today differ from the Empire State Building?
   _____
   _____

Reading Comprehension *Grade 5*

Pronouns

> **REMEMBER**
> A **pronoun** takes the place of a noun or nouns in a sentence.
> Singular pronouns are **I**, **me**, **you**, **he**, **him**, **she**, **her**, and **it**.
> Plural pronouns are **we**, **us**, **you**, **they**, and **them**.
> The word or group of words that a pronoun refers to is called the
> **antecedent**. A pronoun must agree with its antecedent.
>
> **EXAMPLES**
>
> | Antecedent | Pronoun |
> |---|---|
> | **Bryan** works in a restaurant. | **He** is a waiter. |
> | Many **people** work at the restaurant. | **They** perform all kinds of jobs. |

Directions Write the antecedent of each underlined pronoun.

1. Carson's Restaurant is a popular place in our town.
   <u>It</u> has been in business for many years. _____

2. Penny has worked at Carson's for three years.
   <u>She</u> is a college student. _____

3. Tia has never worked in a restaurant.
   <u>She</u> is learning quickly. _____

4. Mr. Carson likes to hire young people.
   <u>He</u> says they are hard workers. _____

5. Jacob clears dishes off each table and takes
   <u>them</u> to Allen and Sam. _____

6. Allen and Sam wash dishes at the restaurant.
   <u>They</u> are a hard-working team. _____

7. I don't like the dishwater.
   <u>It</u> makes my hands wrinkle. _____

8. Bryan has had several jobs at Carson's.
   <u>He</u> used to clear the tables. _____

9. Lucia and Anton bake fresh bread and rolls.
   <u>They</u> also help make desserts. _____

10. Anton makes delicious carrot cake.
    <u>It</u> makes my mouth water! _____

Vacation Variation

Directions Complete the analogies using words from the box below.

city	arctic	planet	building	solar system
café	restaurant	Rome	courtyard	history

1. EARTH is to _____ as SUN is to STAR.

2. CHINA is to COUNTRY as PHILADELPHIA is to _____.

3. HOT is to TROPIC as COLD is to _____.

4. TREE is to FOREST as PLANET is to _____.

E-mail Me!

Directions Read the e-mail message. Use words from the box above to fill in the blanks.

Dear Eli,

I am drinking coffee in a tiny sidewalk **(5)** _____ in sunny Rome. It is a beautiful **(6)** _____ where many musicians and artists work. The tall **(7)** _____ you see in the background is hundreds of years old. There are ruins nearby that are thousands of years old! **(8)** _____ was once a great empire that ruled large parts of the world. I have learned so much about its **(9)** _____!

Tonight we are going to eat at a family-owned **(10)** _____ that is located in a small **(11)** _____, or open area. The food is supposed to be delicious, and I like eating outside. I'll write later.

Love,

Rachel

Estimating Mixed Numbers

REMEMBER
When subtracting or adding **mixed numbers**, an **estimate** is all you need. The fractions may not be important. To estimate, round each mixed number to a whole number before you add or subtract. To round a mixed number you can:

- round to the **lower** whole number by dropping the fraction.
  $7\frac{1}{4}$ rounds to 7 $12\frac{7}{8}$ rounds to 12

- round to the **higher** whole number.
  $7\frac{1}{4}$ rounds to 8 $12\frac{7}{8}$ rounds to 13

- round to the **nearest** whole number. This is the most accurate approach.
  $7\frac{1}{4}$ rounds to 7 $12\frac{7}{8}$ rounds to 13

Directions Estimate an answer for each addition or subtraction problem. Round each mixed number to the nearest whole number. The first one is done for you.

1. $8\frac{1}{4}$ Estimate: 8
    $+3\frac{3}{4}$ $+4$
    $11\frac{4}{4}$ or 12   12

2. $6\frac{2}{3}$ Estimate:
    $+3\frac{1}{8}$

3. $10\frac{1}{4}$ Estimate:
    $+8\frac{6}{8}$

4. $5\frac{2}{3}$
    $-2\frac{1}{5}$

5. $12\frac{7}{8}$
    $-6\frac{1}{8}$

6. $8\frac{9}{12}$
    $-3\frac{1}{4}$

24 Math *Grade 5*

Pool Proportions

Directions Read each problem. Then complete the page.

1. The Amazon Community Pool is in the shape of a rectangle. The pool is 25 yards long and 15 yards wide. Make a drawing of the pool. Label the length of each side.

2. What is the perimeter of the pool? Show your work.

   _____ yards

3. What is the area of the pool? Show your work.

   _____ square yards

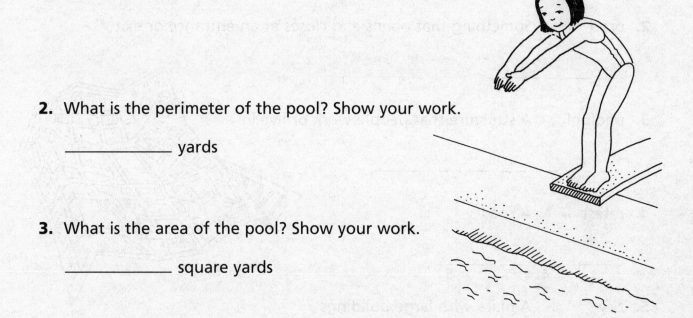

4. The drawing shows a rectangular walkway around the pool that is 2 yards wide. What is the perimeter of the outside edge of the walkway? Show your work.

   _____ yards

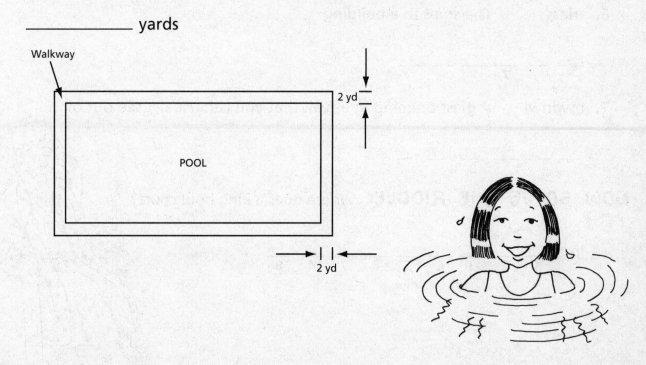

Math *Grade 5*

Around a Building

Directions Read each definition. Unscramble the letters to solve. Write the words on the lines. Then use the numbered letters to solve the riddle.

1. focief   A room in a building where a person works

    __ __ __ __ __ __
             1

2. orod   Something that opens and closes at an entrance or exit

    __ __ __ __
       2

3. undigbli   A structure that people work or live in

    __ __ __ __ __ __ __ __
     3

4. retest   A road

    __ __ __ __ __ __
       4   5

5. ticy   A place with large buildings

    __ __ __ __
         6

6. triass   The steps in a building

    __ __ __ __ __ __
         7

7. owdnwi   A glass opening in a wall that you can usually see out of

    __ __ __ __ __ __
         8

NOW SOLVE THE RIDDLE: Where does a king hold court?

In a __ __ __ __ __ __ __ __
     1 2 3 4 5 6 7 8

26 Word Scramble *Grade 5*

Challenge Yourself: Math

Using a Floor Plan

Lisa's family is moving into a new apartment. The diagram below is called a floor plan. It shows the size of each room.

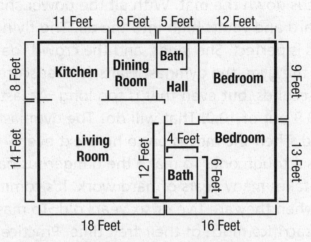

Directions Use the floor plan to answer the questions.

1. What is the perimeter of the living room? (Hint: Include the length of the doorways when calculating the perimeter.)

   _____

2. What is the area of the kitchen?

   _____

3. Lisa's mom is planning to put new carpet in both bedrooms, not including the master bath. How many square feet of carpeting does she need?

   _____

4. Both of the doorways to the dining room are 2 feet wide. Lisa's dad is replacing the baseboards in the dining room. How many feet of baseboard will he need?

   _____

CHAPTER 4
Extraordinary Athletes

The gymnast speeds down the mat. With all the power she has, she jumps on the springboard and vaults high into the air. The flying, twisting somersault that follows is perfect. She lands, and the crowd roars in approval. As cheers fill the gym, the gymnast waits for her score. It takes the judges only a few seconds, but even that is too long. At last the score flashes on the screen: 9.9 out of 10.0. That will do! The gymnast smiles and waves to the crowd. Then she moves on to her next event, the balance beam.

A gymnast's life is a tough one. To make the dangerous handsprings, cartwheels, and tumbles look easy takes many years of hard work. It's common for boys and girls to start their training when they are five or six years old. To master gymnastics, these athletes sacrifice much of their free time. Practice often begins at 5 A.M. and lasts until school starts. Then at 3:30 P.M., it's back to the gym for more practice. To become good enough to compete in meets, gymnasts practice up to six hours a day—every day. When gymnasts show enough skill, power, focus, and grace, they may win a spot on the United States gymnastics team. Every four years, this team goes to the Olympics. This is the meet many gymnasts have been working toward all their lives.

At the Olympics, they will compete against other great gymnasts from around the world. They'll have to perform the most extraordinary routines. If they do well, they may bring home a gold medal.

Extraordinary Athletes

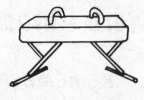

Directions Using what you have just read, answer the questions.

1. Why did the author write this article?
   _____
   _____

2. In the beginning of the article, what happened after the gymnast completed her vault?
   _____
   _____

3. Why could gymnastics be dangerous?
   _____
   _____

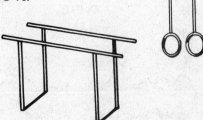

4. Which sentence from the article is an opinion? Circle it.

   A gymnast's life is a tough one.

   Every four years, this team goes to the Olympics.

5. What are the Olympics?
   _____
   _____

6. What does the article mean when it says **they may bring home a gold medal**?
   _____
   _____

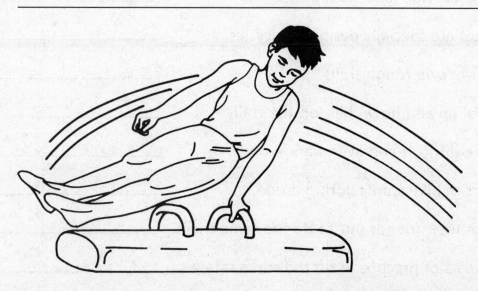

Reading Comprehension Grade 5

Verb Phrases

REMEMBER
An action verb is a word that expresses action, such as **jump, hike,** and **kick**. The verb **be** is a **linking verb**. It connects the subject of a sentence with a noun, pronoun, or adjective. Linking verbs tell what a subject is or is like.
Am, is, and **are** tell about things that are happening now.
Was and **were** tell about things that happened in the past.

A **verb phrase** is a group of words that does the work of a single verb.
In a verb phrase, the **main verb** names the main action.
A **helping verb** comes before the main verb. It helps to show action.

Common Helping Verbs

am	was	did	had	may	could
is	were	have	shall	can	should
are	do	has	will	might	would

Bikes in Action

Directions Draw a line under each verb in the sentences below. Then write if the verb is **action** or **linking** on the line.

1. Grant and Michael ride mountain bikes. _____

2. Mountain bikes are specially designed bicycles. _____

3. The knobby tires grip rough trails. _____

4. The boys pedal up and down hills on the trails. _____

5. Some trails are difficult and even dangerous. _____

6. The boys practice bike stunts during a ride. _____

7. The boys warn their friends not to do the same tricks. _____

8. It has taken years of practice to do the tricks safely. _____

Sports Shorts

Directions Read each clue. Then choose words from the box below to solve.

| Arena | champions | ballplayer | congratulate |
| batter | trophy | referee | uniforms |

1. If you are up to bat in baseball, you are the _____.
2. A prize you get for winning a game is a _____.
3. A person who enforces the rules in a sporting match is a _____.
4. A member of a baseball team is a _____.

Play-by-Play

Directions Read the play-by-play call. Use words from the box above to fill in the blanks.

Welcome back to this exciting game here at the Gund (5) _____. We are watching the final game between the Stars and the Knights. The winners will go home with the state basketball (6) _____. Alex McGraw, who many say is the greatest (7) _____ in high school history, is ready to throw the ball in, so here it goes. The Stars, wearing the red (8) _____, need two points to win the game. McGraw passes to Thompson, who turns and shoots. It's off the rim. The Knights grab it. There's the buzzer! The game is over and the Knights have won. The Stars are being great sports! They are shaking hands to (9) _____ all the players on the Knights' team. It looks like the Knights are the (10) _____!

Adding Mixed Numbers

REMEMBER
Follow these steps to add mixed numbers.
- Add the fractions, using a common denominator.
- Then add the whole numbers.
- Reduce to lowest terms if needed.

EXAMPLES

$$2\tfrac{3}{5}$$
$$+\,1\tfrac{1}{5}$$
$$\overline{3\tfrac{4}{5}}$$

$$5\tfrac{1}{4} = 5\tfrac{2}{8}$$
$$+\,3\tfrac{7}{8} = 3\tfrac{7}{8}$$
$$\overline{8\tfrac{9}{8} = 9\tfrac{1}{8}}$$

Directions Add the mixed numbers. Reduce to lowest terms if needed.

1. $3\tfrac{2}{5}$
   $+\,2\tfrac{1}{5}$

2. $9\tfrac{5}{8}$
   $+\,3\tfrac{2}{8}$

3. $12\tfrac{7}{16}$
   $+\,8\tfrac{8}{16}$

4. $4\tfrac{3}{4}$
   $+\,3\tfrac{1}{4}$

5. $3\tfrac{2}{4}$
   $+\,1\tfrac{3}{8}$

6. $7\tfrac{3}{4}$
   $+\,4\tfrac{5}{16}$

7. $5\tfrac{3}{4}$
   $+\,2\tfrac{2}{3}$

8. $11\tfrac{1}{2}$
   $+\,6\tfrac{2}{3}$

Hiking for Miles

Directions Juan and Karen are going on a hiking trip. Their starting point and direction of travel are shown on the map below. Use the map to answer the questions.

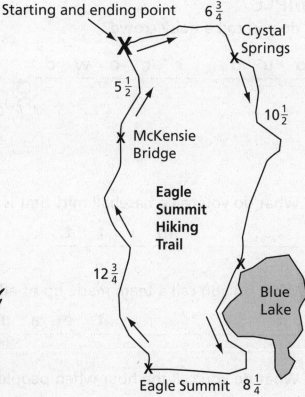

1. Juan and Karen first look at the map to estimate the total distance of the hiking trip. What is their estimate? _____

2. At the starting point, how far are Juan and Karen from Blue Lake? _____

3. Going over Eagle Summit, how far is McKensie Bridge from Blue Lake? _____

4. The hikers plan to fish at a creek about 3 miles beyond Eagle Summit. Place an X on the map to show the fishing spot.

5. If they complete the hike, what is the total trail distance Juan and Karen will travel? _____

6. Juan and Karen hike about 2 miles each hour. Estimate about how many hours it would take them to hike the whole loop. _____

Secret Sports

Directions Use each clue to solve the riddles. Hint: Think about words that rhyme with those in dark print.

EXAMPLE
What do you call a noisy **crowd**?

l o u d   c r o w d

1. What do you call a baseball **mitt** that is the right size?

   ___ ___ ___   m i t t

2. What do you call a **team** made up of all the best players?

   ___ ___ ___ ___ ___   t e a m

3. What do you call the hour when people **climb** a mountain?

   c l i m b   ___ ___ ___ ___

4. What do you call a **group** of people who like to play basketball?

   ___ ___ ___ ___   g r o u p

5. What do you call the money you pay to **ski** on a slope?

   s k i   ___ ___ ___

6. What would you call a sprint around the **bases**?

   b a s e   ___ ___ ___ ___ ___

Challenge Yourself: Reading

A Summer Hike

Every summer my family hikes up Long Pass Trail. We begin before sunrise. Although my brother complains that it's not light yet, I love the dim early morning light. I appreciate how pretty nature looks just before dawn. As we walk up the mountain, the trail gets tougher. The path grows steadily steeper, and rocks and tree roots make climbing more difficult. By lunch we reach flatter ground. Delicate pink and white flowers beg us to notice them as they sway in the wind. We enjoy as much natural beauty as we can. Then we begin the challenging trip back down the trail. By the time we reach the bottom, we're exhausted but happy!

1. Why did the author write this passage?

    A. to describe a family trip

    B. to give information about hiking

    C. to explain why hiking is difficult

    D. to describe the colors of flowers

2. What is the setting of the passage?

    A. a campground   B. Long Pass Trail

    C. the author's house   D. a beach

3. Which of the following is a fact?

    A. Every summer my family hikes up Long Pass Trail.

    B. Nature looks most beautiful before sunrise.

    C. The steeper the trail, the more fun the hike.

    D. Everyone should go out and appreciate nature's beauty.

4. How does the narrator feel about nature?

    A. She thinks it's beautiful.

    B. She thinks it's scary.

    C. She doesn't care about it.

    D. She thinks it's dangerous.

CHAPTER 5
The Spirit of Jazz

Music is said to be a kind of language because it "speaks" to us in different ways. Music talks through sound, rhythm, and melody. Like all languages, music is a form of expression. One musical form (the only one created in America) is jazz. Perhaps more so than any other kind of music, jazz speaks the language of emotions.

Jazz grew naturally from the old rhythms of West African music. For a long time, African Americans had molded these rhythms into church songs and folk music. In the early 1900s, musicians began inventing and changing the music while they were playing it. This was the spirit of jazz—creating music to show a feeling.

The heart of jazz music is in New Orleans. Early bands often had two sections. The rhythm section was made up of the piano, drums, and bass players. These musicians played music that was already written. In the other section were the horns, clarinet, and trombone. As the rhythm section played, the other players would pipe in one by one. Taking their cues from each other, the bands made up melodies right on the spot. Audiences were thrilled by the music. It was different and exciting. The moods of the music, which changed from happy and light to sad and dark, were personal to each musician.

It wasn't long before jazz music spread north. New Orleans musicians like Louis Armstrong, who played the cornet and trumpet, and Jelly Roll Morton, a piano player, moved to Chicago. Soon it became the second major center for jazz. In Chicago, more musicians joined the jazz movement, including such greats as saxophonist Bud Freeman and clarinet player Benny Goodman. Singers began joining jazz bands, too. Bessie Smith, Ella Fitzgerald, and Billie Holiday excited audiences with their beautiful, emotional voices.

Jazz music has always been a changing art form. As new musicians and singers came onto the scene, they brought new and different sounds with them. But that's the spirit of jazz—making music that's personal, emotional, and thrilling for audiences.

The Spirit of Jazz

Directions Using what you have just read, answer the questions.

1. What is this article mostly about?

2. What does the article mean when it says that jazz speaks the language of emotions?

3. Where in the United States did jazz begin?

4. What two sections of instruments made up early jazz bands?

5. What happened in a jazz tune that audiences found different and exciting?

6. Is jazz music different from the music you listen to? Tell how.

Reading Comprehension *Grade 5*

Verb Tenses: Past, Present, and Future

REMEMBER
The **tense** of a verb tells when an action takes place.
A **present tense verb** shows action that is happening now.
A **past tense verb** shows action that has already happened.
A **future tense verb** shows action that will happen.

EXAMPLES
Present tense: Kids of all ages **act** in movies.
Past tense: Last year, Louisa **acted** in her school's musical.
Future tense: Next week, she **will act** in an adventure movie.

Directions Draw a line under each verb. Then write if it is **present**, **past**, or **future tense** on the line.

1. An ad in the paper announces tryouts for a musical. _____

2. My friend Jamie told me about the ad earlier in the day. _____

3. After school, Jamie will call for more information. _____

4. The musical will feature well-known singers. _____

5. Jamie sang in a television commercial last summer. _____

6. He takes singing lessons at the Center for the Arts. _____

7. Jamie's aunt studies singing in New York. _____

8. She will appear in a Broadway play next September. _____

Same Songs

Directions Synonyms are words that are similar in meaning. Match each word from the box with its synonym.

| rap | recite | speaker | rejoice |
| yell | beat | music | speechless |

1. shout _____

2. announcer _____

3. celebrate _____

4. tune _____

5. say _____

6. rhythm _____

Musical Roots

Directions Read the essay. Use words from the box above to fill in the blanks.

There are many different kinds of (7) _____. Some music historians believe that all music began with drums in Africa. Popular music includes rock, country, pop, blues, and (8) _____. One form of the blues is known as scat. Singers say short phrases like "doop" and "bop." Others (9) _____ poetry in their music. They say the rhymes to the (10) _____ of the drum. Some music is so powerful, it leaves listeners (11) _____.

Vocabulary Grade 5

Subtracting Mixed Numbers

REMEMBER
Follow these steps to subtract mixed numbers.
- Subtract the fractions, using a common denominator.
- Then subtract the whole numbers.
- Reduce to lowest terms if needed.

EXAMPLES

$$4\tfrac{5}{6} - 1\tfrac{4}{6} = 3\tfrac{1}{6}$$

$$9\tfrac{2}{3} = 9\tfrac{4}{6} \quad -2\tfrac{1}{6} = 2\tfrac{1}{6} \quad 7\tfrac{3}{6} = 7\tfrac{1}{2}$$

Directions Subtract the mixed numbers. Reduce to lowest terms if needed.

1. $4\tfrac{3}{4} - 1\tfrac{2}{4}$

2. $7\tfrac{6}{8} - 3\tfrac{1}{8}$

3. $15\tfrac{7}{16} - 4\tfrac{3}{16}$

4. $10\tfrac{7}{8} - 5\tfrac{1}{8}$

5. $5\tfrac{3}{4} - 2\tfrac{3}{8}$

6. $9\tfrac{2}{3} - 7\tfrac{1}{5}$

7. $6\tfrac{3}{4} - 1\tfrac{2}{3}$

8. $12\tfrac{5}{6} - 7\tfrac{1}{4}$

Penny Probability

Directions A penny has two sides: the heads side and the tails side. Use this information to answer the questions.

1. If you flip a penny one time, what is the probability that the penny will land heads up? Write the probability as a fraction. _____

2. Suppose you flip a penny one time and it lands heads up. If you flip the same penny a second time, what is the probability that this second flip will also land heads up? Give a reason for your answer.

   _____
   _____

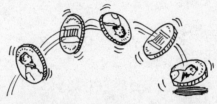

3. If you flip a penny 30 times, how many "heads up" flips are you most likely to get? Give a reason for your answer.

   _____
   _____

4. If you flip two pennies at the same time, what is the probability that both will land heads up? Give a reason for your answer.

   _____
   _____

5. Flip a penny ten times. Record the way the penny lands in the chart. Write **H** for "heads" and **T** for "tails."

1	2	3	4	5	6	7	8	9	10

6. What did the experiment above show?

   _____
   _____

Math Grade 5

Song Search

Directions Find the words listed in the box. Then circle them in the puzzle. The words can be hidden across, down, or diagonally.

rap	beat	chant	country	rhythm
jazz	tune	music	singer	instrument

```
c  i  t  u  e  j  r  l  o  m
e  h  a  x  e  r  a  q  v  u
d  b  a  t  k  y  p  z  w  s
i  z  p  n  p  o  l  s  z  i
c  o  u  n  t  r  y  i  c  c
i  t  e  i  a  t  s  n  a  v
n  u  h  s  e  h  b  g  m  k
i  n  s  t  r  u  m  e  n  t
f  e  i  e  h  s  a  r  a  u
b  l  r  h  y  t  h  m  g  t
```

42 Word Search *Grade 5*

Challenge Yourself: Math

Number Cubes

Directions A six-sided cube with numbers 1–6 is rolled. Use this information to answer the questions.

1. If you roll the cube one time, what is the probability the cube will land on 4? Write the probability as a fraction. _____

2. Suppose you roll the cube twice. On the first roll, the cube lands on 6. What is the probability the second roll will also be a 6? Write the probability as a fraction. _____

    Is the probability of rolling 1 twice in a row greater, less, or the same as rolling 6 twice in a row? Explain your answer.

    _____
    _____
    _____

3. If you roll one cube 18 times, how many times do you predict the cube would land on 3? _____

4. Roll a six-sided cube 12 times. Record the numbers on which the cube lands in the chart.

1	2	3	4	5	6	7	8	9	10	11	12

What did the experiment above show?

_____
_____

CHAPTER 6
Why Animals Can't Talk

In the beginning, animals roamed the wilderness. Then the Iroquois people came. Wolf showed the Iroquois the best hunting grounds. Beaver built dams so they had lakes to fish in. Sheep showed them green meadows. Dog hunted with them and watched over their children.

The Iroquois people lived well and their numbers grew. Soon, too many were hunting in Wolf's hunting grounds. Too many were fishing in Beaver's lake. Beaver was a little upset, Wolf was furious, and even Sheep felt crowded. The animals called a meeting and did not invite the people.

"I am terribly, horribly angry," said Wolf. "I think we animals should eat all the people."

"'That's too dangerous," said Beaver. "Let's do something sneaky. Let's knock down their houses in the middle of the night."

"That's too underhanded," said Sheep in a gentle voice. "Let's just lead them away to new meadows."

"I like the people," said Dog. "I don't mind sharing with them."

Just then, a burst of light flooded the forest. The Great Spirit arrived. "I have heard you animals talking," he boomed. "Because of this nasty plot, none of you will ever speak again."

"But wait," pleaded Dog. "Not all of us wanted to hurt the people." Sheep agreed with Dog.

The Great Spirit sighed, "I suppose you are right. I will still take away your language, but Dog will not be punished as harshly. Dog, you will still understand the people's language, and you will continue to live with them."

The Great Spirit turned to Sheep. "You wanted to lead the people away. For this, you will give up your wool to clothe the people. You did not want to harm them, so they will not harm you."

"Beaver," The Great Spirit frowned. "You wanted to knock down the peoples' houses. For this, the people will think you are a pest and will drive you out of the lakes."

"Wolf," The Great Spirit shook his head sadly, "you wanted to eat the people. For this, you will be a sworn enemy of the people. They will chase you out of their lands." The Great Spirit clapped his hands, and it was so. To this day, no animal can speak the language of the people.

Why Animals Can't Talk

Directions Using what you have just read, answer the questions.

1. How did Sheep help the Iroquois people?
   _____
   _____

2. What happened after the animals helped the people?
   _____
   _____
   _____

3. Why was Beaver upset that so many people were fishing in his lake?
   _____
   _____
   _____

4. How are Dog and Sheep alike?
   _____
   _____

5. Why did The Great Spirit take away the animals' language?
   _____
   _____

6. What special gift did Dog get to keep?
   _____
   _____

Reading Comprehension *Grade 5*

Predicate Adjectives

REMEMBER
A **predicate adjective** is an adjective that follows a linking verb and describes the subject of a sentence. A linking verb joins the subject of a sentence with a predicate adjective. The most common linking verb is **to be** in its various forms: **am, is, are, was, were, being,** and **been**. Verbs such as **seem, feel, look,** and **grow** are also used as linking verbs.

EXAMPLES
The monkeys are **playful**.
The cheetahs look **fierce**.

Directions Complete each sentence below by adding a predicate adjective. Then underline the word that the predicate adjective describes in the sentence.

1. The African lions at the Wildlife Park were _____.

2. I felt _____ as I watched the rhinoceros in their outdoor setting.

3. The crocodiles near the water looked _____.

4. The expression on the gorilla's face seemed _____.

5. Many of the animals in the park are _____.

6. A Masai giraffe looked _____ among his companions.

7. All the animals seemed _____ in the Wildlife Park.

8. A trip to the park is always _____.

Wildlife Comparisons

Directions Complete the analogies using words from the box below.

joey	parrots	elephant	octopus
cobra	chicken	mosquito	horse

1. HORSE is to FOAL as KANGAROO is to _____.
2. RAT is to ANIMAL as _____ is to INSECT.
3. STARFISH is to FIVE as _____ is to EIGHT.
4. BEAR is to WOODS as _____ is to FARM.
5. FISH are to SCALES as _____ are to FEATHERS.
6. SLITHER is to SNAKE as GALLOP is to _____.

All About the Animals

Directions Read the letter. Use words from the box above to fill in the blanks.

Dear Ms. Dean,

　I want to thank you for showing us around the animal sanctuary. I liked learning how you're trying to protect wild animals. In the aquarium, we saw an (7) _____ using its eight strong arms, or tentacles, to swim. There were so many beautiful (8) _____ flying around the bird room. One of them said hello to me! In the jungle area, we saw a big gray (9) _____. He was spraying water on himself with his trunk. There was also a (10) _____ slithering around. I learned a lot!

Thank you,

Josh Taylor

Vocabulary Grade 5

Working with Decimals

REMEMBER
Fractions are one way to represent a whole. **Decimals** represent a whole, too. The fraction $\frac{1}{10}$ can be written as the decimal **0.1**. The first digit to the right of a decimal point is the **10ths**.

A whole number followed by a decimal is called a **mixed decimal**.

EXAMPLES

Decimal	In words	Meaning
0.3	three tenths	3 parts out of 10
1.4	one and four tenths	1 plus 0.4

Decimal Work

Directions Write each number using decimals.

1. two tenths _____
2. $\frac{1}{10}$ _____
3. five tenths _____
4. $\frac{3}{10}$ _____
5. seven tenths _____
6. $\frac{9}{10}$ _____

Directions Write each number using words.

7. 0.4 _____
8. 2.3 _____
9. 0.5 _____
10. 8.1 _____

48 Math Grade 5

Measure the Distance

REMEMBER
A **centimeter ruler** uses tenths. The centimeter ruler is divided into **centimeters (cm)** and **millimeters (mm)**. Each millimeter is one-tenth of a centimeter.

$$1 \text{ mm} = 0.1 \text{ cm}$$

Ruler Rules

Directions Use the centimeter ruler to complete the page.

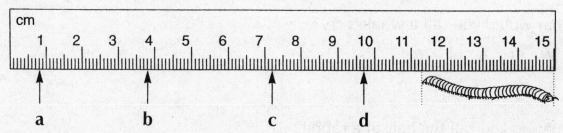

1. Write each distance in centimeters. Don't forget the decimal point!

   A. _____ cm   B. _____ cm

   C. _____ cm   D. _____ cm

2. Mark the following distances on the ruler above.

   A. 1.8 cm   B. 4.9 cm   C. 7.5 cm   D. 12.2 cm

3. How long is the centipede? _____ cm
   (Careful! The centipede is at the wrong end of the ruler!)

4. Read each length. Find something that is about that long. Write the item on the line.

   A. 9 mm _____

   B. 3.0 cm _____

   C. 8.5 cm _____

   D. 12.7 cm _____

Math Grade 5   **49**

Animal Imposters

Directions Homonyms are words that sound alike but have different spellings and meanings. Solve these riddles using homonym word pairs. Hint: Think about the words in dark print.

EXAMPLE: What do you call a bird in an angry, **foul** mood?

A <u>f</u> <u>o</u> <u>w</u> <u>l</u>   <u>f</u> <u>o</u> <u>u</u> <u>l</u>

1. What do you call a wild hog that tells stories that **bore** the listeners?

   A __ __ __ __   <u>b</u> <u>o</u> <u>r</u> <u>e</u>

2. What would you call a **whale's** cry?

   A <u>w</u> <u>h</u> <u>a</u> <u>l</u> <u>e</u> __ __ __ __ __

3. What do you call the **hair** of a rabbit?

   __ __ __ __ __   <u>h</u> <u>a</u> <u>i</u> <u>r</u>

4. What do you call a **horse** that's lost his voice?

   A __ __ __ __ __ __   <u>h</u> <u>o</u> <u>r</u> <u>s</u> <u>e</u>

5. How did the **deer** begin a letter?

   __ __ __ __   <u>d</u> <u>e</u> <u>e</u> <u>r</u>

6. What is a story about a tiger's **tail**?

   A <u>t</u> <u>a</u> <u>i</u> <u>l</u> __ __ __ __ __

Challenge Yourself: Reading

Pet Sitting

Cara sometimes takes care of her neighbor's poodle. She also feeds her aunt's cat when her aunt is away from home. Now Cara wants to start a pet-sitting business. She will use the money she earns from pet sitting to pay for her school trip.

The first thing Cara must do is find a way to let people know about her pet-sitting business. She is making flyers to pass out to her parents' friends. Cara's neighbor and aunt are going to recommend her to their friends.

Next, Cara needs to make a plan so that she will have time to take care of all the animals and still get her homework done. She will get up an hour earlier than usual so she has time to visit the animals each morning. After school, she will walk the dogs and spend time brushing the cats and playing with them. She will ride her bike to her clients' homes instead of walking. Setting up a business is a lot of work, but Cara is determined to be successful.

Directions Circle the letter of the best answer.

1. What is this story mostly about?

    A. Cara takes care of a neighbor's poodle.

    B. Cara wants to go on a school trip.

    C. Cara is starting a pet-sitting business.

    D. Cara makes flyers to advertise her business.

2. How many times each day does Cara plan to visit the animals?

    A. once a day   B. three times a day

    C. twice a day   D. four times a day

3. Which of the following words best describes Cara?

    A. shy   B. careless

    C. determined   D. bossy

4. Choose the best answer to complete the following analogy.

    POODLE is to DOG as _____ is to REPTILE.

    A. cat   B. mosquito

    C. horse   D. snake

Challenge Yourself: Reading Grade 5

CHAPTER 7
Scotland's Sea Monster

Is there a sea monster in Loch Ness, one of Scotland's biggest lakes? No one knows for sure. Some say yes and call the creature the "Loch Ness Monster." Some say no. Many pictures of the monster exist. All the pictures are too blurry to prove anything, though.

How would a monster from the sea get into a lake anyway? One theory takes into account the history of Scotland's land and oceans. Thousands of years ago, Loch Ness was not a lake, but an arm of the North Sea. Its salty waters were full of sea creatures. Some of these creatures, such as sea cows and giant squids, grew to be very large. If we saw them today, we might call them monsters.

After the Ice Age, the land in Northern Scotland began to rise and the seas began to shrink back. The land beneath the North Sea rose and trapped some of the seawater inside Scotland. This water became a lake. Over time, springs and streams fed the lake with freshwater. Today, we know this lake as Loch Ness, home of the mythical Loch Ness Monster.

Loch Ness is about 24 miles long and 1 mile wide. It is in the Scottish Highlands, where it never freezes. The lake runs along a large fault, or crack in Earth's surface. The fault causes small earthquakes that shake the waters of the lake every few years. It also makes the lake very deep. It's over 700-feet deep in some spots!

Could sea monsters have been trapped when the lake was created? Could these creatures have adapted to the freshwater of the lake? Because the water is so deep and so muddy, people may never know. We do know that our Earth has changed over time, turning part of the sea into Loch Ness. The monsters of Loch Ness could really be nothing more than freshwater sea creatures.

Scotland's Sea Monster

Directions Using what you have just read, answer the questions.

1. What is this article mainly about?

   _____
   _____

2. Why do some people believe there is a Loch Ness Monster?

   _____
   _____

3. Why do some people not believe the Loch Ness Monster exists?

   _____
   _____

4. What happened to Northern Scotland after the Ice Age?

   _____
   _____

5. Why did the author write this article?

   _____
   _____

6. Do you think there is some kind of monster that lives in Loch Ness? Why or why not?

   _____
   _____
   _____

Conjunctions

REMEMBER
Conjunctions are words that join other words or groups of words in a sentence. The most common conjunctions are **and**, **but**, **or**, and **nor**. Conjunctions can join two or more nouns, verbs, adjectives, or adverbs in a sentence. They can also connect two separate thoughts or ideas in a sentence.

EXAMPLES
Joining Nouns: Rachel **and** Maria gave a report about jellyfish.
Joining Verbs: We wrote **and** proofread the report.
Joining Adjectives: The report was long **but** interesting.
Joining Adverbs: Rachel checked our report carefully **and** completely.
Joining Complete Thoughts: We might add to our report, **or** we might choose another topic.

Creature Conjunctions

Directions Draw a line under the conjunction in each sentence.

1. Jellyfish are simple but fascinating animals.
2. They have bell-shaped bodies and trailing tentacles.
3. Some jellyfish are up to 20 feet long and 10 feet wide.
4. Nerve cells tell a jellyfish if it is headed up or down.
5. Jellyfish eat tiny floating animals, fish, and other jellies.
6. All jellies sting, but not all jellyfish are poisonous to humans.
7. Corals and sea anemones are related to jellyfish.
8. Jellyfish can live in warm or cold water.
9. They are pushed through the water by wind and currents.
10. Jellyfish look like slippery blobs, but they are graceful creatures.

Landscape Likeness

Directions Complete the analogies using words from the box below.

> plateau   continent   cornfield   volcano
> fields   glaciers   mountain   ocean

1. HILL is to _____ as VALLEY is to CANYON.

2. WHEAT is to WHEATFIELD as CORN is to _____.

3. FORESTS are to TREES as _____ are to GRASS.

4. EARTH is to PLANET as NORTH AMERICA is to _____.

5. FLOOD is to RIVER as ERUPT is to _____.

6. PEAK is to MOUNTAIN as FLAT is to _____.

Erie Essay

Directions Read the essay. Use words from the box above to fill in the blanks. You may need to add an -s to some words.

My name is Ryan. I live near Lake Erie. The Great Lakes, like Lake Erie, were carved out of the earth when the (7) _____ melted and receded north. The lakes are fresh water, not salt water like the (8) _____. Although we have hills, they are nothing like the Rocky (9) _____. In winter, snow turns the grassy (10) _____ white.

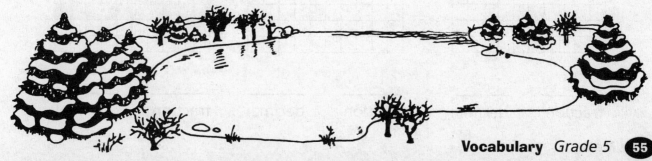

Vocabulary Grade 5   55

In the Hundredths Place

REMEMBER
You already know about the hundredths place. You use it when you write cents. The amount $1.65 has a 5 in the hundredths place: 5 stands for 5¢. The cents place is the hundredths place because each penny is one-hundredth of a dollar. The hundredths place is the second place to the right of the decimal point.

EXAMPLES

Decimal	In words	Meaning
0.05	five hundredths	5 parts out of 100
2.07	two and seven hundredths	2 plus 0.07
3.99	three and ninety-nine hundredths	3 plus 0.99

Write It Out

Directions Write each number as a decimal.

1. nine hundredths _____
2. $\frac{7}{100}$ _____
3. seven hundredths _____
4. $\frac{65}{100}$ _____
5. eighty-one hundredths _____
6. $\frac{83}{100}$ _____

Directions Write a fraction and a decimal that shows the shaded part of each grid.

7.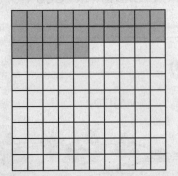

   _____   _____
   fraction decimal

8.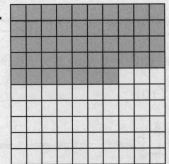

   _____   _____
   fraction decimal

9.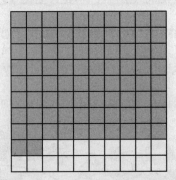

   _____   _____
   fraction decimal

Fish Finds

Directions Follow the directions to complete the chart.

Fish	Actual Cost (A)	Rounded to nearest dime (B)	Rounded to next higher dollar (C)
tetra	$1.89		
goldfish	$2.08		
shark	$4.87		
guppy	$1.39		
catfish	$2.78		
beta	$3.19		

1. Round each amount to the nearest dime. Write the rounded amounts in column **B** on the chart.

2. Estimate the total cost by adding the amounts in column **B**.

   Estimated total cost: _____

3. Round each amount to the next higher dollar. Write the rounded amounts in column **C** on the chart.

4. Estimate the total cost by adding the amounts in column **C**.

   Estimated total cost: _____

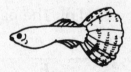

5. Add the prices in column **A** to find the actual total cost.

   Actual total cost: _____

6. Circle the estimate that is most accurate, or gives an estimate closest to the actual total cost.

   rounding to the nearest dime

   or

   rounding to the next higher dollar

Math Grade 5

Sea Scramble

Directions Read each clue. Unscramble the letters to find the hidden words. Write the words on the lines. Then use the numbered letters to solve the riddle.

1. naoce   A large body of water

   __ __ __ __ __
          1

2. spotuco   An animal with eight legs that lives in the ocean

   __ __ __ __ __ __ __
              2

3. last   The mineral in ocean water

   __ __ __ __
        3

4. rhaks   An ocean animal with sharp teeth

   __ __ __ __ __
         4

5. flyilsehj   An ocean animal that has long tentacles

   __ __ __ __ __ __ __ __ __
              5

6. sligl   The part of a fish that helps it breathe under water

   __ __ __ __ __
     6

7. lehaw   A large ocean mammal

   __ __ __ __ __
      7

NOW SOLVE THE RIDDLE: Which ocean animal sings in a band?

__ __ __ __ __ __ __ __
 1 2 3 1 4 5 6 2 7

58 Word Scramble *Grade 5*

Challenge Yourself: Math

In the Thousandths Place

REMEMBER
The second digit to the right of the decimal is in the hundredths place. The third digit to the right of the decimal is in the thousandths place. The number 0.148 has an 8 in the thousandths place. 8 stands for 8 out of a thousand; 0.148 stands for 148 out of a thousand. 0.148 can be written in words as one hundred forty-eight thousandths.

EXAMPLES

Decimal	In Words	Meaning
0.003	three thousandths	3 parts out of 1,000
6.007	six and seven thousandths	6 plus 0.007
0.429	four hundred twenty-nine thousandths	429 parts out of 1,000
11.035	eleven and thirty-five thousandths	11 plus 0.035

Directions Write the words as a decimal or each fraction in words.

1. eight thousandths _____
2. 5/1,000 _____
3. three and nineteen thousandths _____
4. 7/1,000 _____
5. twenty-two and six-hundred forty-seven thousandths _____
6. 92/1,000 _____

Directions Write each decimal in words.

7. 0.414 _____
8. 5.027 _____
9. 3.194 _____

CHAPTER 8

Bai-Nien

In China, and in cities like San Francisco where Chinese-Americans live, the streets fill with crowds on the first day of the lunar (or moon) calendar. Everyone shouts and sings, "Bai-Nien!" (BEE nee-EN) "Happy New Year!"

Chinese New Year is a cheerful holiday. For three days, workers stop working and students stop studying. The government closes its doors. Everyone stops what they are doing to give thanks for the ending year. It is time to welcome in the New Year with a festival.

The party begins on New Year's Eve, when families gather for a tasty meal. They eat dumplings, rice cakes, and fruit. They decorate their homes with red cloth, paper flowers, and candles. In Chinese tradition, each year is represented by one of twelve different animals. For instance, 2002 was the Year of the Horse. At the meal, families have pictures or puppets of that year's special animal on display. At midnight, just as the New Year arrives, everyone rushes outside to see fireworks light up the night sky.

On New Year's Day, many people visit their temples to give thanks to the gods. Then the parades begin. The day turns into a rich mosaic of bright colors, singing children, and dancing animals. People join in the Dragon Dance. Friends and relatives greet one another on the streets and at home. They give each other gifts and say "Bai-Nien!"

Chinese New Year used to last two full weeks. In most places, the holiday doesn't last that long anymore. However, at the end of the two weeks, people still take part in the Lantern Festival. A 2,000-year-old custom, the Lantern Festival is a nighttime parade under a full moon. Everyone carries colorful paper lanterns that they've made or bought. Leading the parade is the Golden Dragon, made of bamboo and covered with cloth or paper. In China, the Golden Dragon, like all dragons, stands for strength and goodness. It is a sign of hope for the year ahead. "Bai-Nien!"

Bai-Nien

Directions Using what you have just read, answer the questions.

1. What is Bai-Nien?
   _____
   _____

2. What closes during Bai-Nien?
   _____
   _____
   _____

3. What might the people give thanks for?
   _____
   _____
   _____

4. Which sentence is an opinion? Circle it.

   In Chinese tradition, each year is represented by one of twelve different animals.

   Chinese New Year is a cheerful holiday.

5. Which events would you like to participate in during Bai-Nien? Name two.
   _____
   _____
   _____

6. How is Bai-Nien like the American New Year? How is it different?
   _____
   _____
   _____

Subjects: Simple and Complete

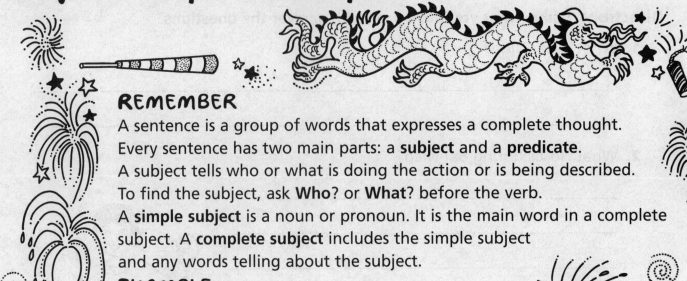

REMEMBER
A sentence is a group of words that expresses a complete thought. Every sentence has two main parts: a **subject** and a **predicate**. A subject tells who or what is doing the action or is being described. To find the subject, ask **Who?** or **What?** before the verb.
A **simple subject** is a noun or pronoun. It is the main word in a complete subject. A **complete subject** includes the simple subject and any words telling about the subject.

EXAMPLE
      Complete subject
   (Hoang Anh's <u>family</u> /) escaped from Vietnam.
         ↑
    Simple subject

Subjects and Verbs

Directions Draw a slash between the complete subject and the verb in each sentence. Then underline the simple subject in the sentence. The first one is done for you.

1. The <u>journey</u> by boat / was difficult for Hoang's family.

2. The small, overcrowded boat faced many hardships.

3. Hoang's family lived in a refugee camp for more than a year.

4. Many other people from Vietnam lived in the camp.

5. Conditions in the camp were poor.

6. A church in Oregon helped Hoang's family come to the United States.

7. The people in the church found a place for the family to live.

8. Hoang's father became a crab fisherman.

9. Hoang's family learned about the American holiday called Independence Day.

10. Everyone liked to celebrate because they were free, too.

62 Language *Grade 5*

Holiday Picnic

Directions Look at each clue. Choose words from the box below that match each clue.

> barbecue   plate   salad   lemonade
> watermelon   jelly   picnic

1. grill _____
2. fruit _____
3. basket _____
4. pitcher _____

Packing for the Picnic

Directions Read the steps below. Use words from the box above to fill in the blanks.

Packing for a July Fourth **(5)** _____

Step 1: Mix 2 cups of cut apples, 2 cups of sliced strawberries, 2 peeled oranges, and 1 scooped **(6)** _____ to make a fresh fruit **(7)** _____.

Step 2: Toast 5 English muffins, cover each with 1 teaspoon of butter and strawberry **(8)** _____.

Step 3: Pack the hot dogs in the cooler for the **(9)** _____.

Step 4: Don't forget the pitcher of **(10)** _____ to drink!

Vocabulary *Grade 5*   63

Comparing Decimals

REMEMBER
When you compare decimals, you see which number is **greater than** or **less than** the other. To compare decimals, give each decimal the same number of digits. Then compare the digits from left to right.

EXAMPLES

Which is **greater than**: 0.4 or 0.38?

0.40 ← Place a 0 in the 100ths place.
0.38

0.4 > 0.38 because 40 > 30.

Which is **less than**: 0.2 or 0.17?

0.20 ← Place a 0 in the 100ths place.
0.17

0.17 < 0.20 because 17 < 20.

Greater or Less?

Directions Write < or > between each pair to compare the numbers.

1. 0.15 _____ 0.2
2. 0.3 _____ 0.28
3. 0.57 _____ 0.5
4. 0.07 _____ 0.1
5. 2.4 _____ 2.08
6. 3.8 _____ 3.94

Directions Shade each grid the amount indicated. Then write < or > in the box to compare the numbers.

7. 0.09 ☐ 0.1

8. 0.52 ☐ 0.4

64 Math Grade 5

Track Times

Directions Use the information and chart below to answer the questions.

The River City School had a track meet. The track coach listed the top five best times in the girl's 100-yard dash on a chart. The fastest time run in the 100-yard dash at this meet was 11.15 seconds.

Name	Time (in seconds)
Tyine	12.19
Alicia	11.5
Stacey	11.87
Keisha	12.6
Marni	11.34

1. Write the River City runners' names in order of finish.

1st: _____ (fastest time)

2nd: _____

3rd: _____

4th: _____

5th: _____ (slowest time)

2. By how much did each runner go over the record time of 11.15 seconds?

Tyine: _____

Alicia: _____

Stacey: _____

Keisha: _____

Marni: _____

3. What was the median score of the River City runners? _____

4. What was the range of the River City runners? _____

5. What was the average time of the River City runners? _____

Math Grade 5

Holiday Puzzle

Directions Use words from this chapter to help solve the puzzle.

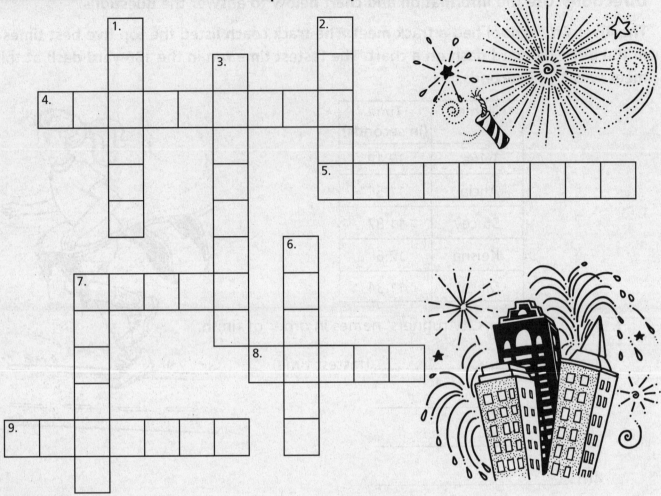

ACROSS
4. Colorful explosions in the night sky
5. To honor a special time or holiday
7. A happy gathering of people
8. Presents
9. People shout this during Chinese New Year.

DOWN
1. People marching with floats and bands to honor something special
2. A song is this
3. Thanksgiving and Labor Day are these
6. A group of people who are related
7. A meal eaten outside

Challenge Yourself: Reading

May Day

May Day is celebrated on May 1 each year. To follow an old American tradition, on May Day you can place a small basket of flowers or treats on a neighbor's doorstep. Then ring the doorbell and try to run away before the door is opened. If you are caught, tradition says you are supposed to get a kiss! Nowadays, not many Americans remember May Day. That's too bad, because it is fun to celebrate.

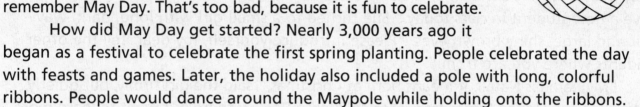

How did May Day get started? Nearly 3,000 years ago it began as a festival to celebrate the first spring planting. People celebrated the day with feasts and games. Later, the holiday also included a pole with long, colorful ribbons. People would dance around the Maypole while holding onto the ribbons.

May Day is now celebrated around the world. In England, children bring flowers to their neighbors and get pennies in return. In France, people tie flowers to the tails of cows and lead the cows in a parade.

1. What do people in France do on May Day that is different from what Americans do on May Day?

    A. In France, people lead cows in parades.

    B. In France, children get pennies on May Day.

    C. In France, children bring flowers to their neighbors.

    D. In France, people tie ribbons around the necks of cows.

2. Which sentence from the selection is an opinion?

    A. May Day is celebrated on May 1 each year.

    B. That's too bad, because it is fun to celebrate.

    C. People celebrated the day with feasts and games.

    D. May Day is now celebrated around the world.

3. What did May Day first celebrate?

    A. the harvest of crops   **B.** the first spring planting

    C. the birth of baby animals   **D.** the coming of summer

CHAPTER 9

A New Student

Arnold had just hit Mindy in the head with a paper airplane when we heard our teacher's footsteps in the hall. Mindy was pretending to be mad. I think she just loved being the center of attention. She would never actually tell on Arnold. I didn't want the attention, so I just tried to stay away from that boy.

When Ms. Baxter came in, the usual hush fell over the classroom. "Ladies and Gentlemen," said Ms. Baxter with a half-smile on her face, "*Mesdames et Messieurs,* (may DAMS ay may SYRS,) *Meine Damen und Herren* (MY na DAM en unt HAIR en), we have a new student in class today." She turned to a small girl with long, dark, wavy hair and large, shiny brown eyes. "Please introduce yourself, my dear. Tell the other students where you are from."

"My name is Natali. It means born at Christmas," said the girl firmly, but quietly. "I come to the United States from Mexico. I learn English there but want to learn more."

"You can see, class, why I am adding *Damas y Caballeros* (DAM as e COB ul YER os) to my usual greeting," said Ms. Baxter cheerily. "That means Ladies and Gentlemen in Spanish. Now get out your writing journals. Everyone will write about his or her first day of school. I want you to include a lot of detail and feeling." She showed Natali to an empty desk near me, Mindy, and Arnold.

Ms. Baxter went to the front of the room. Mindy whispered, "The new girl talks funny." Arnold giggled. Natali turned and stared at Mindy. Mindy giggled again.

"*Tu no hablas Espanol,*" (TU no HAB las ES pon YOL) said Natali in a cold voice. I could tell that meant something like—you don't even speak Spanish. Mindy and Arnold giggled more. Natali stared at them until they stopped giggling and turned away. She was so brave, although I saw her hands tremble as she got out her notebook. I knew right then that I wanted to be her friend!

68 Reading Comprehension *Grade 5*

A New Student

Directions Using what you have just read, answer the questions.

1. Where does this story take place?
   _____
   _____

2. How do you know that Ms. Baxter likes people from other countries?
   _____
   _____

3. How does the writer feel about Arnold? How do you know?
   _____
   _____

4. Why does Mindy tease Natali?
   _____
   _____

5. The story says that Natali was brave. List two things Natali does to show she is brave.
   _____
   _____

6. Do you think the writer of the story will be friends with Natali? Why or why not?
   _____
   _____

Reading Comprehension Grade 5

Sentence Predicates

 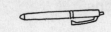  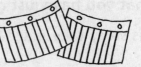 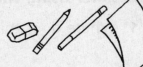

REMEMBER
Every subject has two main parts: the **subject** and the **predicate**. The predicate of a sentence tells what the subject is or does. To find the predicate, ask **What does the subject do?** or **What is the subject like?** The **simple predicate** is a verb or verb phrase that expresses action or a state of being. The **complete predicate** includes the verb and any words that tell about the action or complete its meaning.

EXAMPLE

Complete predicate

Tiffany wants to take a class to learn how to cook.

— Simple predicate

Find the Predicate

Directions Draw a line between the subject and the predicate in each sentence. Then underline the simple predicate. The first one is done for you.

1. The students in the cooking class / <u>prepared</u> a banquet.
2. Their parents attended the banquet.
3. First, the students served a melon and mango salad.
4. The salad had thin slices of ham in it.
5. Everyone enjoyed the salad.
6. The students had prepared a chicken roulade.
7. A roulade is a type of stuffed meat.
8. Steamed asparagus was served with the roulade.
9. The asparagus was flavored with butter and lemon juice.
10. They also served fluffy mashed potatoes.

 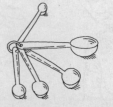

70 Language Grade 5

Time Scramble

Directions Unscramble the letters to make words from the box below.

awhile	current	delay	during	following
gradual	since	sudden	until	

1. wonliglof _____
2. urencrt _____
3. ledya _____
4. luadrag _____

Lost and Found

Directions Read the diary entry. Use words from the box above to fill in the blanks.

Dear Diary:

You will not believe what happened today! Our class took a field trip to the museum. I was having a great time (5) _____ I got separated from my friends. It took (6) _____ before I realized that I was lost. I was walking and thought everyone was (7) _____ me. At least 25 minutes had passed (8) _____ I'd last seen my friends. I knew that (9) _____ that time they could have walked pretty far. I found a security guard and asked for help when, all of a (10) _____, I heard someone call my name. They had found me!

Vocabulary Grade 5

Adding Numbers with Decimals

REMEMBER
You add decimals the same way you add whole numbers. Just remember to line up the decimal points and write the decimal point in the answer.

EXAMPLES

```
  1            1            1
  0.46         0.78         0.87
+ 0.27       + 0.51       + 0.40
  0.73         1.29         1.27
```

Add It Up

Directions Add the decimals.

1.   0.6
   + 0.3

2.   0.5
   + 0.2

3.   0.7
   + 0.1

4.   0.8
   + 0.4

5.   2.3
   + 1.4

6.   3.5
   + 2.7

7.   4.2
   + 3.9

8.   1.65
   + 0.80

9. 2.5 + 1.24 = _____

10. 3.9 + 0.84 = _____

11. 1.83 + 0.7 = _____

72 Math *Grade 5*

Number Sense

Directions Complete the page.

1. Evan saw the following problem in his math book:

    Mary bought several pencils for $0.25 each. She spent $2.50.
    How many pencils did Mary buy?

    Assume that ☐ stands for the number of pencils that Mary bought.

    Circle the equation that is true for ☐.

    A. $0.25 + ☐ = $2.50   **B.** $0.25 − ☐ = $2.50

    C. $0.25 × ☐ = $2.50   **D.** $0.25 ÷ ☐ = $2.50

2. Now solve the equation. How many pencils did Mary buy? _____

3. Evan then saw a similar problem written this way:

    Mary buys pencils for $0.25 each.
    How much will she spend if she buys 7 pencils?

    Assume that ○ stands for the amount Mary spent.

    Circle the equation that is true for ○.

    A. ○ + $0.25 = $7.00   **B.** ○ − $0.25 = $7.00

    C. ○ × $0.25 = $7.00   **D.** ○ ÷ $0.25 = $7.00

4. Now solve the equation. How much will she spend if she buys 7 pencils? _____

5. Evan then saw this third problem:

    Mary spent $3.38 for 13 pencils. What was the price of each pencil?

    Assume that △ stands for the price of each pencil.

    Write an equation that is true for △. Then solve the equation.

Math Grade 5

Student Search

Directions Circle the words listed in the box. The words can be hidden across, down, or diagonally.

class	ruler	science	computer	library
book	reading	teacher	student	cafeteria

```
f  i  r  u  e  c  o  o  s  t
b  c  e  x  e  o  a  q  v  e
c  l  a  s  s  m  p  z  w  a
l  s  d  f  r  p  e  s  z  c
i  c  i  n  e  u  y  i  r  h
b  i  n  o  b  t  b  n  a  e
r  e  g  s  l  e  e  o  m  r
a  n  e  t  z  r  m  r  o  t
r  c  o  r  u  l  e  r  i  k
y  e  s  t  u  d  e  n  t  a
```

Challenge Yourself: Math

Adding Decimals

Directions Add the decimals and write the sum.

1. 0.41 + 0.57 = _____
2. 2.12 + 4.59 = _____
3. 6.202 + 4.131 = _____
4. 6.008 + 2.019 = _____
5. 3.124 + 7.120 = _____
6. 0.303 + 0.921 = _____
7. 21.274 + 0.312 = _____
8. 4.321 + 6.008 = _____
9. 12.003 + 4.207 = _____
10. 6.994 + 3.201 = _____
11. 0.045 + 3.612 = _____
12. 9.217 + 6.495 = _____
13. 7.202 + 8.633 = _____

14. 4.003 + 2.711 = _____
15. 8.735 + 9.204 = _____
16. 14.615 + 4.222 = _____
17. 7.417 + 3.285 = _____
18. 6.221 + 7.413 = _____
19. 8.214 + 3.552 = _____
20. 3.214 + 7.367 = _____
21. 5.204 + 9.003 = _____
22. 3.072 + 8.403 = _____
23. 2.011 + 7.921 = _____
24. 6.319 + 8.224 = _____
25. 0.447 + 3.217 = _____
26. 4.521 + 13.007 = _____

Challenge Yourself: Math Grade 5

CHAPTER 10
The Plant Doctor

The next time you see sweet potatoes or peanuts, think of George Washington Carver. He is the scientist who helped make these foods popular.

Carver was a famous African-American botanist. He was a scientist who studied plants and tried to find ways to use them to improve people's lives.

George Washington Carver was born in 1864. He was a slave on Moses Carver's farm in Diamond Grove, Missouri. After he was born, Confederate soldiers kidnapped George and his mother. George was returned to the Carvers' farm, but his mother was never heard from again. The Carvers raised George and his brother as their own children.

Young Carver wanted to be a scientist. He went to a one-room school for African-American children. He graduated from high school with honors. Because he was African American, most colleges would not accept him. But in 1891, Carver was the first African-American student accepted at Iowa State University.

After he graduated, Carver joined the teaching staff at Iowa State. He was also the first African-American to become a member of the staff there.

In 1896, Tuskegee Institute, a college for African-American students in Alabama, hired Carver to lead their new department of agriculture. He worked as a botanist. Because he was always experimenting with plants, people soon began calling him "the plant doctor."

Carver found that the soil in many parts of the South had been ruined by years of growing only cotton. Cotton takes nutrients from the soil, but does not return any. Carver found that crops like peanuts and sweet potatoes improved the soil.

Carver began a "school on wheels." This traveling classroom taught farmers how to improve the soil. Farmers soon produced more peanuts and sweet potatoes than people could use. Carver continued his research. He invented 325 products from peanuts and over 100 from sweet potatoes. From these two plants, he made plastics, oils, dyes, medicines, soaps, cereals, powdered milk, and flour.

Carver also loved to paint pictures of large, colorful flowers. Two of his paintings were shown at the 1893 World's Fair in Chicago, Illinois. In 1940, Carver gave all of his savings to the Tuskegee Institute. They built the George Washington Carver Foundation for research in natural science. In 1943, Carver died in Tuskegee, Alabama. He is still honored today as one of America's most important and famous scientists.

The Plant Doctor

Directions Using what you have just read, answer the questions.

1. How would you describe George Washington Carver?

2. Why did people call Carver **the plant doctor**?

3. What did Carver do after he graduated from Iowa State University?

4. What is a botanist?

5. Why do you think Carver started his "school on wheels"?

6. Why did the author write this article?

Reading Comprehension *Grade 5*

Sentences

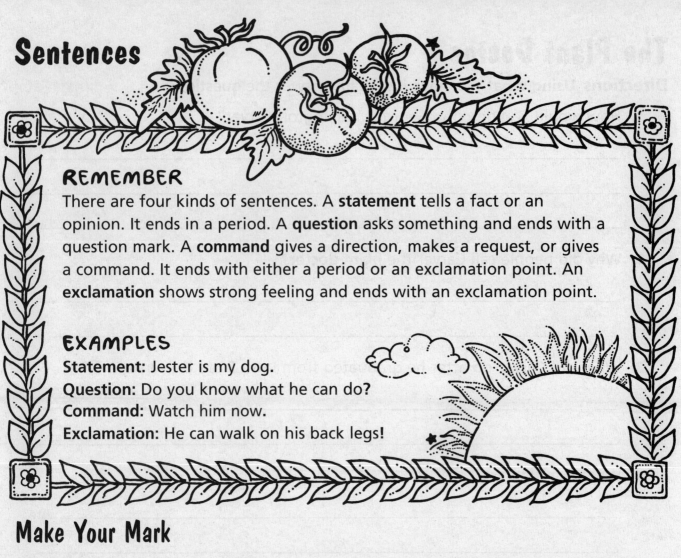

REMEMBER
There are four kinds of sentences. A **statement** tells a fact or an opinion. It ends in a period. A **question** asks something and ends with a question mark. A **command** gives a direction, makes a request, or gives a command. It ends with either a period or an exclamation point. An **exclamation** shows strong feeling and ends with an exclamation point.

EXAMPLES
Statement: Jester is my dog.
Question: Do you know what he can do?
Command: Watch him now.
Exclamation: He can walk on his back legs!

Make Your Mark

Directions Write the most likely end punctuation mark for each sentence. Then write S for a statement, Q for a question, C for a command, or E for an exclamation on the line.

1. I was watering the garden one day _____
2. A squirrel scampered by me _____
3. Do you know what happened next _____
4. Jester streaked after the squirrel _____
5. Jester ran through the garden _____
6. Stop now, Jester _____
7. He crushed my mother's tomato plants _____
8. Will my mother be angry _____
9. I hope not _____

78 Language *Grade 5*

Pick Your Plants

Directions Choose a word from the box below that is most similar to the word listed.

| soil | cinnamon | dogwood | garden | spinach |
| flowers | ivy | vegetable | shrubs | |

1. ginger _____

2. dirt _____

3. bushes _____

Garden Shopping

Directions Read the conversation. Use words from the box above to fill in the blanks.

Clerk: "Welcome to Merry Gardens Nursery. How may I help you?"

Shopper: "I'm looking for a tree and some plants for my garden."

Clerk: "Well, the **(4)** _____ is a very pretty tree. It has beautiful white **(5)** _____ in the spring. Is there anything else you need?"

Shopper: "I'd also like to grow some **(6)** _____ that will creep up the wall!"

Clerk: "Certainly we have different types of climbing plants that I can show you."

Shopper: "I also want to grow tomatoes and lettuce. Do you sell **(7)** _____ plants and seeds?"

Clerk: "If you like lettuce, you might want to try some leafy, green **(8)** _____."

Shopper: "That sounds great!"

Clerk: "I will get all these things for you. It sounds like you are planting quite a **(9)** _____."

Subtracting Decimals

REMEMBER
You subtract decimals the same way you subtract whole numbers. Just remember to line up the decimal points and write the decimal point in the answer.

EXAMPLES

```
     0.94          0 1 6             1 1 3
   − 0.32          1.68              2.36
     ____        − 0.95            − 0.90
     0.62          ____              ____
                   0.73              1.46
```

Take It Away

Directions Subtract the decimals.

1. 0.7   2. 0.8   3. 0.7   4. 1.5
  − 0.2   − 0.4   − 0.3   − 0.8

5. 1.35   6. 2.28   7. 5.12   8. 3.10
  − 0.24   − 0.08   − 2.71   − 1.05

9. 0.47 − 0.21 =   10. 0.94 − 0.63 =   11. 3.5 − 2.27 =

   _____   _____   _____

80 Math Grade 5

Cube Creations

Directions The figure below is made out of 5 cubes. The front- and right-side views of the figure are shown. Use the figures to answer questions 1–3. Then complete the page.

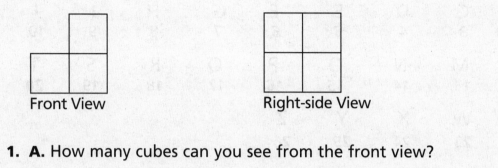

Front View   Right-side View

1. **A.** How many cubes can you see from the front view?   _____

   B. How many cubes can you see from the right view?   _____

2. Think about what the bottom view would look like. How many cubes would you see?   _____

3. On the grids below, draw the two possible bottom views.

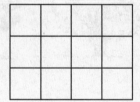   

4. Which of the following figures can be cut out and folded into a cube? Circle them.

A. 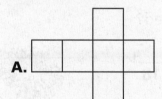   B. 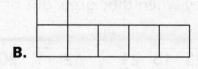   C.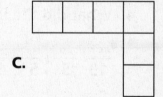

D. 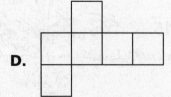   E. 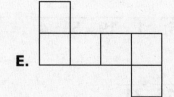   F.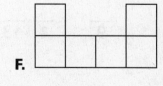

Math *Grade 5*   **81**

Plant Codes

Directions Look at the code. Each number stands for a letter. Write the letters on the lines to solve the riddles.

A	B	C	D	E	F	G	H	I	J
1	2	3	4	5	6	7	8	9	10
K	L	M	N	O	P	Q	R	S	T
11	12	13	14	15	16	17	18	19	20
U	V	W	X	Y	Z				
21	22	23	24	25	26				

1. What kind of root likes to dance?

   ___ ___ ___ ___   ___ ___ ___ ___
    1 20   1 16   18 15 15 20

2. What kind of dance does a tree do?

   ___ ___ ___   ___ ___ ___ ___ ___
   20   8   5   12   9 13   2 15

3. What kind of tree cries?

   ___   ___ ___ ___ ___ ___ ___ ___   ___ ___ ___ ___ ___ ___
    1   23   5   5 16   9 14   7   23   9 12 12 15 23

4. What did the leaves say when they grew out of the seed?

   ___ ___ ___ ___   ___ ___ ___   ___ ___ ___ ___ ___ ___!
   19   5   5   4   25 15 21   12   1 20   5 18

5. What did the pea say when it met the ham?

   ___   ___ ___   ___ ___ ___ ___ ___ ___   ___ ___
    9   1 13   16   5   1 19   5   4   20 15

   ___ ___ ___ ___   ___ ___ ___.
   13   5   5 20   25 15 21

Challenge Yourself: Reading

The Wonderful Sweet Potato

Sweet potatoes have been around for thousands of years. They were grown in the South American country of Peru as early as 750 B.C.E. The Native Americans of Central America were growing and eating sweet potatoes when Columbus arrived in the Americas in 1492. In fact, Columbus brought the first sweet potatoes to Europe.

Today, sweet potatoes are eaten all around the world. In the United States, almost 100,000 acres of sweet potatoes were planted in 2004. The biggest producers of sweet potatoes in the country include North Carolina, California, Mississippi, and Louisiana. However, almost 90 percent of the world's sweet potatoes are grown in Asia.

Sweet potatoes are both healthy and delicious. They are rich in important vitamins, such as vitamins A and C. Cooked sweet potatoes are often eaten plain, as a side dish served with other foods. Mashed sweet potatoes, baked in a pan with marshmallows on top, are quite delicious. Cooked sweet potatoes can also be used in desserts, like pies and pudding. Sweet potatoes can even be eaten raw. They can be cut up and mixed into a salad.

Directions Circle the letter of the best answer.

1. Why do you think the author wrote this article?

    A. to explain how to make a sweet potato dish

    B. to give information about sweet potatoes

    C. to tell about the best ways to eat sweet potatoes

    D. to tell a funny story about sweet potatoes

2. Where are most of the world's sweet potatoes grown?

    A. Peru **B.** Asia

    C. Central America **D.** Louisiana

3. What makes sweet potatoes a healthy food?

    A. They are rich in vitamins A and C.

    B. They can be made into desserts.

    C. They are eaten all around the world.

    D. They are delicious with marshmallows.

READING CHECK-UP
Emil's Garden

Many centuries ago in Turkey, a man named Emil worked in his garden, tilling and weeding the soil. From time to time, he stopped and leaned on his rake to gaze at the fine fruits and vegetables he had grown. He admired the deep red pomegranates (POM uh GRAN its) with their tart juice. Emil bent down and tenderly pulled weeds away from the plump, purple eggplants. He could almost taste the mouth-watering dishes he would make for dinner.

Then Emil noticed something peculiar. He stared at a pistachio tree that stood in the corner of his garden and wondered aloud, "What foolishness it is that a tiny nut should grow from such a huge tree!" Emil went back to his labor, puzzling over the ways of nature.

Then Emil spied watermelons growing on long, tangled vines. "Humph, a stem no thicker than my finger feeds that enormous fruit!" he said. "If I had designed the watermelon, it would grow from a tall, sturdy tree."

Just then, the weather turned. As Emil glanced at the sky, a breeze knocked several pistachios from a branch, and they pelted him on the head. "Aha!" he exclaimed, as each nut hit him. "Indeed, I should stick to my work in the garden and not redesign that which nature has planned. If watermelons grew on trees, surely I would be lying on the ground now. What a melon head I am!"

Directions Fill in the bubble next to the correct answer.

1. This story is **mostly** about:
   - (A) a garden with fruits and vegetables.
   - (B) the differences between watermelons and pistachios.
   - (C) a man who learns something about himself and nature.
   - (D) ingredients for cooking a dinner.

2. Which detail from the story **best** supports the main idea?
   - (A) If watermelons grew on trees, surely I would be lying on the ground now.
   - (B) He stopped and leaned on his rake to gaze at the fine fruits and vegetables he had grown.
   - (C) He stared at a pistachio tree that stood in the corner of his garden.
   - (D) "What a melon head I am!"

3. Which word **best** describes Emil?
   - (A) curious
   - (B) sociable
   - (C) stubborn
   - (D) uninterested

4. From the story, you can conclude that Emil:
   - (A) was hurt by the pistachios hitting him on the head.
   - (B) will not try to redesign nature.
   - (C) will never garden again.
   - (D) does not like to plant flowers.

READING CHECK-UP
A Rapid Ride

"Rock ahead," Pedro shouted as he clung to the ropes along the side of the rubber raft. He and several other children from the City Wide Youth Group were on their first white-water rafting trip. Pedro thought rafting was more exciting than a visit to any amusement park.

"Pedro, you're leaning over too far," called his friend Maria from the back of the raft. "Pull yourself in, Pedro," shouted Alyce, the river guide. But Pedro wasn't listening. He was too caught up in the excitement.

In an instant, the raft struck the rock. The raft lifted high in the air, then came crashing back down into the fast moving water. Pedro was nowhere to be found.

"Where's Pedro?" cried Maria, desperately searching the water.

"I see him up ahead. He's all right," Alyce shouted. The raft quickly caught up with Pedro. "Maria, hold out your paddle. Quickly!" ordered Alyce. Pedro grabbed the paddle and Alyce pulled him into the raft. She then issued a firm warning. "Next time, Pedro, listen to my commands."

"Don't worry," said a shaken Pedro. "It won't happen again." Pedro resumed paddling. This time though, he chose a seat in the middle of the raft.

Directions Fill in the bubble next to the correct answer.

5. Which of these happened first?
   - (A) Alyce issued a firm warning.
   - (B) Pedro chose a new seat.
   - (C) The raft struck a rock.
   - (D) Pedro leaned over too far.

6. From the story, you can tell
   - (A) Pedro had never been rafting before.
   - (B) Alyce and Pedro are close friends.
   - (C) Maria was an experienced swimmer.
   - (D) Alyce was a careless guide.

7. How did Maria feel when she could not see her friend?
   - (A) angry
   - (B) fearful
   - (C) relieved
   - (D) disappointed

8. The next time Pedro goes rafting, he will probably
   - (A) guide the raft himself.
   - (B) pay close attention to the guide's commands.
   - (C) take a seat at the front of the raft.
   - (D) lean over the side of the raft.

READING CHECK-UP
Daniel Boone

Daniel Boone is one of America's most famous pioneers. His love for exploring the land made him the subject of many popular legends.

Boone was born in Pennsylvania on November 1, 1734. Unlike most boys his age, young Daniel Boone was encouraged by his parents to hunt and explore the outdoors on his own. The Boones knew this would best prepare their son for frontier life.

In 1750, Boone moved with his family to North Carolina. He spent most of his time trapping, hunting, or exploring. He joined the army for a brief period in 1755 before deciding that army life was not for him. He decided to travel, visiting Florida and later settling in Kentucky.

Boone enjoyed the Kentucky area so much that he settled there in the 1770s. He helped build three settlements, one of which was named Boonesborough after him. Over time, the arrival of many settlers made Boone hungry for more space. He headed to Missouri in 1799 where he lived out his remaining years.

Directions Fill in the bubble next to the correct answer.

9. Which of these is an opinion?
   - A Daniel Boone is one of America's most famous pioneers.
   - B Boone spent most of his time trapping, hunting, and exploring.
   - C Boone moved with his family to North Carolina in 1750.
   - D He joined the army for a brief period in 1755.

10. Unlike most boys his age, young Daniel Boone
    - A was not allowed to hunt.
    - B was allowed to explore the outdoors on his own.
    - C was the subject of many popular legends.
    - D was allowed to join the army.

11. Why did Boone move to Missouri?
    - A He wanted to explore the area.
    - B His family was living there.
    - C He wanted more open space.
    - D He wanted to start a new settlement.

12. The author wrote this article to
    - A convince readers to learn more about pioneers.
    - B describe America in the 1700s.
    - C entertain readers with a legend about a pioneer.
    - D inform readers about the life of a famous American.

READING CHECK-UP
Starfish

When is a fish not really a fish? When it is a starfish. Starfish are marine invertebrates, animals without backbones, like jellyfish. Real fish have backbones.

Starfish have five or more arms and are often shaped like a star. Some starfish, the Sunflower Star for example, have twenty-four arms. Others can have up to forty-four. Starfish, such as the Bat Star, can grow back arms that they have lost.

Like real fish, starfish come in many colors. For example, the Cushion Star is colored brown, orange, red, or yellow. The Short Spined Starfish is purple, red, yellow, and white. Starfish vary in size, too. Some are just under an inch, but the Sunflower Starfish can grow to be twenty-six inches wide.

Unlike fish, starfish can move slowly in any direction without turning. They use hundreds of tiny tube feet with suction cups on the ends. These suction cups help them crawl and stick to rocks, plants, and coral. Starfish can move their arms in any direction so that they can wedge themselves into tiny places. Fish use their fins to help them move around. Starfish don't have fins.

Starfish skin is rough and leathery, and often has spines for protection. Fish usually have scales that protect their delicate skin. Both types of animals live together in the sea.

Directions Fill in the bubble next to the correct answer.

13. All starfish have
    - (A) forty-four arms.
    - (B) five or more arms.
    - (C) twenty-four arms.
    - (D) five or fewer arms.

14. Both fish and starfish
    - (A) move using tube feet.
    - (B) move using fins.
    - (C) come in many colors.
    - (D) have scaly skin.

15. Because they have no fins, starfish probably
    - (A) can't move sideways.
    - (B) catch insects at the surface.
    - (C) spend their lives out of water.
    - (D) don't swim as fast as fish.

16. Which is NOT true about starfish?
    - (A) Starfish do not live near fish.
    - (B) Starfish vary in size.
    - (C) Starfish can hide in tiny places.
    - (D) Starfish do not have backbones.

Stop!   Number Correct: _____ out of 16

MATH CHECK-UP
Directions Answer each question.

1. Which of these is **not** the same as 7,650?
   - (A) seven thousand six hundred fifty
   - (B) 7,000 + 600 + 50
   - (C) 7 thousands, 6 hundreds, 5 ones
   - (D) 7 thousands, 6 hundreds, 5 tens

2. What is the value of 7 in the number 670,540?
   - (A) 700
   - (B) 70
   - (C) 7,000
   - (D) 70,000

3. Which of these is $\frac{8}{3}$ expressed as a mixed number?
   - (A) $5\frac{1}{8}$
   - (B) $2\frac{1}{3}$
   - (C) $5\frac{1}{3}$
   - (D) $2\frac{2}{3}$

4. Tia has 2 roses, 4 tulips, and 9 daisies. What is the ratio of tulips to daises?
   - (A) 4 to 9
   - (B) 2 to 4
   - (C) 2 to 9
   - (D) 9 to 4

5. Which number is less than 3,583 and rounds to 4,000 when rounded to the nearest thousand?
   - (A) 3,621
   - (B) 3,541
   - (C) 3,259
   - (D) 3,499

6. What number is more than 2,450 but less than 2,540?
   - (A) 4,250
   - (B) 4,520
   - (C) 2,550
   - (D) 2,504

7. What number can be placed in each box to make the number sentence true?

   15 + ☐ + ☐ + ☐ = 30
   - (A) 4
   - (B) 3
   - (C) 5
   - (D) 10

8. Rob put 6 boxes on each shelf. There are 11 shelves. Which number sentence shows the number in all?
   - (A) 6 + 11 = ☐
   - (B) 6 − 11 = ☐
   - (C) 6 × 11 = ☐
   - (D) 6 ÷ 11 = ☐

MATH CHECK-UP
Directions Answer each question.

9. 585 + 926 + 874 = ☐
   - Ⓐ 2,271
   - Ⓑ 2,385
   - Ⓒ 2,285
   - Ⓓ 1,885

10. 75,617 − 45,920 = ☐
    - Ⓐ 30,797
    - Ⓑ 29,790
    - Ⓒ 29,697
    - Ⓓ 29,317

11. $ 72.83
    + $ 18.45
    - Ⓐ 80.28
    - Ⓑ 81.28
    - Ⓒ 82.28
    - Ⓓ 91.28

12. $25.75 − $6.89 = ☐
    - Ⓐ $18.86
    - Ⓑ $19.67
    - Ⓒ $19.86
    - Ⓓ $19.84

13. 9,480 × 7 = ☐
    - Ⓐ 65,560
    - Ⓑ 65,867
    - Ⓒ 66,360
    - Ⓓ 63,830

14. 8,541 ÷ 3 = ☐
    - Ⓐ 2,110
    - Ⓑ 2,847
    - Ⓒ 2,517
    - Ⓓ 2,633

15. $\frac{3}{5} + \frac{2}{5} =$ ☐
    - Ⓐ $\frac{1}{5}$
    - Ⓑ 1
    - Ⓒ $\frac{2}{3}$
    - Ⓓ $\frac{4}{5}$

16. $1\frac{3}{4} - \frac{1}{2} =$ ☐
    - Ⓐ $\frac{1}{4}$
    - Ⓑ $1\frac{2}{4}$
    - Ⓒ 1
    - Ⓓ $1\frac{1}{4}$

17. 0.7 + 0.33 = ☐
    - Ⓐ 0.40
    - Ⓑ 1.30
    - Ⓒ 1.03
    - Ⓓ 7.33

18. 3.45
    − 0.40
    - Ⓐ 3.05
    - Ⓑ 3.41
    - Ⓒ 3.85
    - Ⓓ 3.49

Math Check-Up Grade 5

MATH CHECK-UP
Directions Answer each question.

19. Which figure has more than one right angle?
    - Ⓐ □
    - Ⓑ ○
    - Ⓒ △
    - Ⓓ ⬡

The graph shows the snowfall amounts for four months. Study the graph. Use the graph to answer questions 20 and 21.

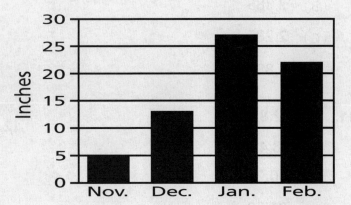

20. Which month had about 22 inches of snow?
    - Ⓐ November
    - Ⓑ December
    - Ⓒ January
    - Ⓓ February

21. About how many more inches of snow fell in December than in November?
    - Ⓐ 5 inches
    - Ⓑ 10 inches
    - Ⓒ 8 inches
    - Ⓓ 3 inches

22. What time will it be in 45 minutes?

    - Ⓐ 11:15
    - Ⓑ 11:25
    - Ⓒ 10:20
    - Ⓓ 11:30

23. Shawn, Mike, Kevin, and Kyle are brothers. Kevin is 10 years younger than Mike. Shawn is twice as old as Kevin. Mike is 16. Kyle is two years younger than Shawn. How old is Kyle?
    - Ⓐ 10
    - Ⓑ 14
    - Ⓒ 18
    - Ⓓ 12

24. What is the perimeter of this shape?

    2 cm | 6 cm

    - Ⓐ 12 cm
    - Ⓑ 8 cm
    - Ⓒ 4 cm
    - Ⓓ 16 cm

25. Which of these shapes is most like a cylinder?
    - Ⓐ a drinking glass
    - Ⓑ a cereal box
    - Ⓒ a book
    - Ⓓ the moon

Stop! Number Correct: _____ out of 25

Congratulations!

_____
(name)

has completed *Summer Counts!*

Good job!

Have a good school year.

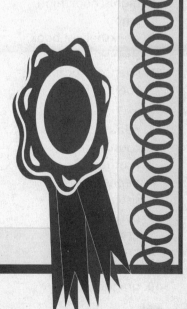

ANSWER KEY

Page 5
1. The girls are twins and look alike. Zinsa has a silver bracelet, and Zinhoue has feathers in her hair.
2. To save her sister's life.
3. Possible answer: They don't like the law. They marry the girls.
4. Possible answer: Althea was jealous because everyone loved the twins.
5. Possible answers: They help each other. They love each other.
6. Possible answer: They didn't agree with the law.

Page 6
1. grandparents: grand, parents
2. golf course: golf, course
3. grandfather: grand, father; newspaper: news, paper
4. teenager: teen, ager
5. bookstore: book, store
6. post office: post, office
7. sister-in-law: sister, in, law
8. high school: high, school
9. bookshelves: book, shelves
10. science fiction: science, fiction

Page 7
1. relative
2. cousin
3. brother
4. nephew
5. sister
6. parents
7. grandmother
8. brother
9. marriage
10. grandchildren

Page 8
1. $\frac{3}{4}$
2. $\frac{6}{9}$ or $\frac{2}{3}$
3. $\frac{2}{6}$ or $\frac{1}{3}$
4. $\frac{6}{10}$ or $\frac{3}{5}$
5. 4; $\frac{5}{6}$
6. 2; $\frac{7}{8}$
7. 4; $\frac{1}{6}$
8. 6; $\frac{1}{10}$

Page 9
1. A. $\frac{25}{100}$ or $\frac{1}{4}$
   B. $\frac{10}{100}$ or $\frac{1}{10}$
   C. $\frac{5}{100}$ or $\frac{1}{20}$
2. A. 25¢ + 10¢ + 5¢ = 40¢
   B. $\frac{40}{100} = \frac{4}{10} = \frac{2}{5}$
3. Possible answers:
   $\frac{5}{20} + \frac{6}{20} + \frac{7}{20} = \frac{18}{20}$ or $\frac{9}{10}$;
   25¢ + 30¢ + 35¢ = 90¢

Page 10
1. aunt
2. grandmother
3. brother
4. grandfather
5. cousin
6. uncle
7. sister
8. parents

Page 11
1. C
2. C
3. A
4. B

Page 13
1. Gutzon Borglum was a sculptor.
2. He loved art and America.
3. Possible answer: He thought the presidents were better subjects.
4. A team of over 100 men began to carve the mountain.
5. George Washington, Thomas Jefferson, Abraham Lincoln, and Theodore Roosevelt
6. It took about 14 years to carve the faces.

Page 14
1. monument
2. carving
3. portrait
4. artist
5. son
6. sculptor
7. heroes
8. memorial
9. sculptures
10. attraction

Page 15
1. doctor
2. director
3. singer
4. students
5. governor
6. artist; sculptor
7. captain
8. police
9. director
10. act

Page 16
1. A. $4\frac{1}{2}$
   B. four and one-half
2. A. $3\frac{1}{3}$
   B. three and one-third
3. A. $2\frac{3}{4}$
   B. two and three-fourths
4. A. $5\frac{1}{3}$
   B. five and one-third

Page 17
Equations may vary.
1. 3 quarts
2. A. $2\frac{1}{4}$
   B. $1\frac{1}{8}$
3. A. 4
   B. 3
   C. $1\frac{1}{2}$
4. A. 8
   B. 6
   C. 3

Page 18
Across
2. teacher
5. police
6. artist
7. sculptor
9. actor

Down
1. dentist
3. director
4. painter
8. doctor

Page 19
1. $\frac{3}{4}$ cups
2. 4 cups
3. 150 cookies
4. sugar – $4\frac{1}{2}$ cups
   brown sugar – 6 cups
   butter – 6 cups
   eggs – 12 eggs
   vanilla extract – 6 teaspoons
   flour – $13\frac{1}{2}$ cups
   baking soda – 6 teaspoons
   salt – 6 teaspoons

Page 21
1. It is about building the Empire State Building.
2. Possible answers: Tall buildings had to be narrower at the top. They had to have light and fresh air.
3. Possible answer: Codes are rules you must follow to make a building safe.
4. It took one year and 45 days to build.
5. The Sears Tower was built after the Empire State Building.
6. Answers may vary. Accept reasonable responses.

Page 22
1. Carson's Restaurant
2. Penny
3. Tia
4. Mr. Carson

5. dishes
6. Allen and Sam
7. dishwater
8. Bryan
9. Lucia and Anton
10. carrot cake

Page 23
1. planet
2. city
3. arctic
4. solar system
5. café
6. city
7. building
8. Rome
9. history
10. restaurant
11. courtyard

Page 24
1. $11\frac{4}{4}$ or 12; $8+4=12$
2. $9\frac{19}{24}$; $7+3=10$
3. $18\frac{8}{8}$ or 19; $10+9=19$
4. $3\frac{7}{15}$; $6-2=4$
5. $6\frac{6}{8}$ or $6\frac{3}{4}$; $13-6=7$
6. $5\frac{6}{12}$ or $5\frac{1}{2}$; $9-3=6$

Page 25
1.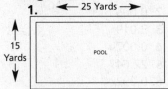
2. 80
3. 375
4. 96

Page 26
1. office
2. door
3. building
4. street
5. city
6. stairs
7. window

Riddle: courtyard

Page 27
1. 64 feet
2. 88 square feet
3. 292 square feet
4. 24 feet

Page 29
1. Possible answer: To tell how hard gymnasts work.
2. The crowd roared in approval.
3. Answers will vary. Accept reasonable responses.
4. A gymnast's life is a tough one.
5. Possible answer: The Olympics are meets many gymnasts have been working toward all their lives.
6. Possible answer: The winner gets a gold medal.

Page 30
1. ride; action
2. are; linking
3. grip; action
4. pedal; action
5. are; linking
6. practice; action
7. warn; action
8. has taken; linking; do; linking

Page 31
1. batter
2. trophy
3. referee
4. ballplayer
5. Arena
6. trophy
7. ballplayer
8. uniforms
9. congratulate
10. champions

Page 32
1. $5\frac{3}{5}$
2. $12\frac{7}{8}$
3. $20\frac{15}{16}$
4. $7\frac{4}{4}$ or 8
5. $4\frac{7}{8}$
6. $11\frac{17}{16}$ or $12\frac{1}{16}$
7. $7\frac{17}{12}$ or $8\frac{5}{12}$
8. $17\frac{7}{6}$ or $18\frac{1}{6}$

Page 33
1. About 45 miles
2. $17\frac{1}{4}$ miles
3. 21 miles
4. An X should be placed about $\frac{1}{4}$ the distance from Eagle Summit to McKensie Bridge.
5. $43\frac{3}{4}$ miles
6. about 22 hours

Page 34
1. fit mitt
2. dream team
3. climb time
4. hoop group
5. ski fee
6. base race

Page 35
1. A
2. B
3. A
4. A

Page 37
1. It is about jazz music.
2. Possible answer: Musicians played jazz music to show their feelings.
3. Jazz began in New Orleans.
4. The rhythm section and the horn, clarinet, and trombone sections.
5. Possible answer: The musicians would take cues from each other and make up music on the spot.
6. Answers will vary. Accept reasonable responses.

Page 38
1. announces; present
2. told; past
3. will call; future
4. will feature; future
5. sang; past
6. takes; present
7. studies; present
8. will appear; future

Page 39
1. yell
2. speaker
3. rejoice
4. music
5. recite
6. beat
7. music
8. rap
9. recite
10. beat
11. speechless

Page 40
1. $3\frac{1}{4}$
2. $4\frac{5}{8}$
3. $11\frac{4}{16}$ or $11\frac{1}{4}$
4. $5\frac{6}{8}$ or $5\frac{3}{4}$
5. $3\frac{3}{8}$
6. $2\frac{7}{15}$
7. $5\frac{1}{12}$
8. $5\frac{7}{12}$

Page 41
1. $\frac{1}{2}$
2. $\frac{1}{2}$; Each time you flip a penny, the probability of getting heads is $\frac{1}{2}$.
3. 15 heads up flips because $\frac{1}{2}$ of 30 is 15.
4. $\frac{1}{4}$; Flipping two pennies gives 4 possible outcomes: 2H, 1H and 1T, 1T and 1H, and 2T. 1 in 4 outcomes are 2H.
5–6. Answers will vary. Accept reasonable responses.

Page 42

Page 43
1. $\frac{1}{6}$
2. $\frac{1}{6}$; It is the same.
3. Estimates will vary.
4. Charts will vary. Answers will vary. Accept reasonable responses.

Page 45
1. Sheep led them to green meadows.
2. The Iroquois people took over the lands.
3. Possible answer: They were disturbing his home.
4. Possible answer: Dog and Sheep did not want to harm the people.
5. Possible answer: The Great Spirit wanted to punish them for planning against the people.
6. Dog could still understand the people.

Page 46
1–8. Predicates will vary. Accept reasonable responses.
1. lions
2. I
3. crocodiles
4. expression
5. animals
6. giraffe
7. animals
8. trip

Page 47
1. joey
2. mosquito
3. octopus
4. chicken
5. parrots
6. horse
7. octopus
8. parrots
9. elephant
10. cobra

Page 48
1. 0.2
2. 0.1
3. 0.5
4. 0.3
5. 0.7
6. 0.9
7. four tenths
8. two and three-tenths
9. five tenths
10. eight and one-tenth

Page 49
1. A. 0.8
   B. 3.8
   C. 7.2
   D. 9.7
2. Should show correct distances on ruler.
3. 3.7 cm
4. A–D. Answers will vary. Accept reasonable responses.

Page 50
1. boar bore
2. whale wail
3. hare hair
4. hoarse horse
5. dear deer
6. tail tale

Page 51
1. C
2. C
3. C
4. D

Page 53
1. It is about Scotland's Loch Ness and the Loch Ness monster.
2. Possible answer: They have seen the monster or pictures of it.
3. Possible answer: They have never seen the monster. The pictures they saw were blurry.
4. The land began to rise and the seas began to shrink.
5. The author wanted to tell about how Loch Ness was formed and the legend of the Loch Ness Monster.
6. Answers will vary. Accept reasonable responses.

Page 54
1. but
2. and
3. and
4. or
5. and
6. but
7. and
8. or
9. and
10. but

Page 55
1. mountain
2. cornfield
3. fields
4. continent
5. volcano
6. plateau
7. glaciers
8. ocean
9. Mountains
10. fields

Page 56
1. 0.09
2. 0.07
3. 0.07
4. 0.65
5. 0.81
6. 0.83
7. $\frac{25}{100}$ or $\frac{1}{4}$; 0.25
8. $\frac{47}{100}$; 0.47
9. $\frac{82}{100}$; 0.82

Page 57
1. tetra: $1.90
   goldfish: $2.10
   shark: $4.90
   guppy: $1.40
   catfish: $2.80
   beta: $3.20
2. $16.30
3. tetra: $2.00
   goldfish: $3.00
   shark: $5.00
   guppy: $2.00
   catfish: $3.00
   beta: $4.00
4. $19.00
5. $16.20
6. rounding to the nearest dime

Page 58
1. ocean
2. octopus
3. salt
4. shark
5. jellyfish
6. gills
7. whale

Riddle: A starfish

Page 59
1. 0.008
2. 0.005
3. 3.019
4. 0.007
5. 22.647
6. 0.092
7. four hundred fourteen thousands
8. five and twenty-seven thousands
9. three and one hundred ninety-four thousands

Page 61
1. Bai-Nien is the Chinese New Year.
2. Possible answers: The government, offices, and schools close.
3. Answers will vary. Accept reasonable responses.
4. Chinese New Year is a cheerful holiday.

5. Answers will vary. Accept reasonable responses.
6. Possible answers: Both celebrations begin on New Year's Eve. Bai-Nien lasts for three days.

Page 62
1. The journey by boat /
2. The small, overcrowded boat /
3. Hoang's family /
4. Many other people from Vietnam /
5. Conditions in the camp /
6. A church in Oregon /
7. The people in the church /
8. Hoang's father /
9. Hoang's family /
10. Everyone /

Page 63
1. barbecue
2. watermelon
3. picnic
4. lemonade
5. picnic
6. watermelon
7. salad
8. jelly
9. barbecue
10. lemonade

Page 64
1. <
2. >
3. >
4. <
5. >
6. <
7. 9 squares shaded; 1 rectangle shaded; 0.09 < 0.1
8. 52 squares shaded; 4 rectangles shaded; 0.52 > 0.4

Page 65
1. 1st: Marni
   2nd: Alicia
   3rd: Stacey
   4th: Tyine
   5th: Keisha
2. 1.04; 0.35; 0.72; 1.45; 0.19
3. 11.87
4. 1.26
5. 11.9

Page 66
Across
4. fireworks
5. celebrate
7. party
8. gifts
9. Bai-Nien

Down
1. parade
2. music
3. holidays
6. family
7. picnic

Page 67
1. A
2. B
3. B

Page 69
1. The story takes place in a classroom.
2. Possible answer: Ms. Baxter greets them in many languages.
3. Possible answer: The writer doesn't like him. She tries to stay away from him.
4. Possible answer: Mindy teases Natali because she speaks differently.
5. Possible answers: Natali spoke to the class even though her English was not perfect. She stared at Mindy and Arnold.
6. Possible answer: Yes. She says she wants to be her friend.

Page 70
1. The students in the cooking class / prepared
2. Their parents / attended
3. First, the students / served
4. The salad / had
5. Everyone / enjoyed
6. The students / had prepared
7. A roulade / is
8. Steamed asparagus / was served
9. The asparagus / was flavored
10. They / also served

Page 71
1. following
2. current
3. delay
4. gradual
5. until
6. awhile
7. following
8. since
9. during
10. sudden

Page 72
1. 0.9
2. 0.7
3. 0.8
4. 1.2
5. 3.7
6. 6.2
7. 8.1
8. 2.45
9. 3.74
10. 4.74
11. 2.53

Page 73
1. C
2. 10
3. D
4. $1.75
5. 26; Possible equation: 13 × △ = $3.38

Page 74
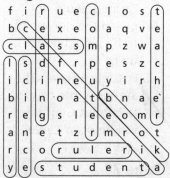

Page 75
1. 0.98
2. 6.71
3. 10.333
4. 8.027
5. 10.244
6. 1.224
7. 21.586
8. 10.329
9. 16.21
10. 10.195
11. 3.657
12. 15.712
13. 15.835
14. 6.714
15. 17.939
16. 18.837
17. 10.702
18. 13.634
19. 11.766
20. 10.581
21. 14.207
22. 11.475
23. 9.932
24. 14.543
25. 3.664
26. 17.528

Page 77
1. Possible answer: Carver was a hard-working man who loved plants.
2. He always experimented with plants.
3. Carver began to teach there.
4. A botanist is someone who studies plants.

5. Possible answer: Farmers could not get to the university, so he would go to them.
6. Possible answer: The author wrote this to tell about George Washington Carver and his work.

Page 78
1. period; S
2. period; S
3. question mark; Q
4. period; S
5. period; S
6. exclamation point; E or C
7. period; S
8. question mark; Q
9. exclamation point; E

Page 79
1. cinnamon
2. soil
3. shrubs
4. dogwood
5. flowers
6. ivy
7. vegetable
8. spinach
9. garden

Page 80
1. 0.5
2. 0.4
3. 0.4
4. 0.7
5. 1.11
6. 2.20
7. 2.41
8. 2.05
9. 0.26
10. 0.31
11. 1.23

Page 81
1. A. 3
   B. 4
2. 3
3.
4. A; D; E

Page 82
1. a tap root
2. the limbo
3. a weeping willow
4. Seed you later!
5. I am peased to meet you.

Page 83
1. B
2. B
3. A

Page 84
1. C
2. A
3. A
4. B

Page 85
5. D
6. A
7. B
8. B

Page 86
9. A
10. B
11. C
12. D

Page 87
13. B
14. C
15. D
16. A

Page 88
1. C
2. D
3. D
4. A
5. B
6. D
7. C
8. C

Page 89
9. B
10. C
11. D
12. A
13. C
14. B
15. B
16. D
17. C
18. A

Page 90
19. A
20. D
21. C
22. B
23. A
24. D
25. A